GRATITUDE IS NOT ENOUGH

The True Story of a Belgian Family Forever Changed by a Band of American WWII Soldiers

Praise For

GRATITUDE IS NOT ENOUGH

"It is often those things you think are worthless that become the most valuable treasures in later life. A few forgotten war relics were the seeds for the Remember Museum 39-45 in Thimister-Clermont, Belgium, and at the same time the beginning of decades of gratitude towards the veterans of the Greatest Generation who were never forgotten by the people they liberated from the Nazi yoke. Hundreds of stories, narratives and fates thus became a memorial to a time that we must never forget. Mathilde and Marcel Schmetz have created a museum with inimitable dedication and attention to detail that is second to none. Through their museum, the M&Ms preserve and honor the stories of these brave boys. In *Gratitude is Not Enough*, Dr. Stein honors the amazing and selfless story of the M&Ms, and preserves the gratitude they so richly deserve."

—Albert Trostorf, Mayor of Langerwehe-Merode, Germany.
WWII historian and author of the book series Lest We Forget ....
concerning the battles of the Hürtgen Forest and surrounding towns.

"*Gratitude Is Not Enough* is NOT what you expect of a military, non-fiction read. When it was described as a love story, I was hesitant. But this is a love story and so much more. I was totally engrossed by the real stories and emotional rollercoaster of this well-researched book.

The M&Ms are an amazing couple, and I am in awe of their dedication to and reverence for the American soldier. Their commitment to showing their appreciation knows no bounds. They've made it their life's mission to ensure the world never forgets what American soldiers sacrificed and achieved in WWII. To work tirelessly and diligently to educate others on the true cost of war is so altruistic, humanitarian, compassionate...I honestly don't know the words to describe the M&Ms.

The author did an amazing job capturing the personal stories of a handful of American soldiers. They weren't unknown, faceless warriors; they were sons, husbands and brothers. I felt a real sense of loss for those who did not survive.

Marcel's life from a young boy to today was simply captivating. The detailed anecdotes shared are eye-opening. Though I'll never be able to fully comprehend the horror of occupation, Marcel's story provides a powerful glimpse.

In spite of the atrocities of occupation and war, I now understand why—on so many levels—this is a love story. My hope is that *Gratitude Is Not Enough* accomplishes its mission to help maintain the legacy of the M&Ms and their Remember 39-45 Museum."

—Jami Lynn Contardi, book reviewer.

GRATITUDE IS NOT ENOUGH

The True Story of a Belgian Family Forever Changed by a Band of American WWII Soldiers

TOM STEIN

Published by

TMS
SERVICES
LLC.

Top cover photo shows Private First Class Robert Blett (right), extending a hand in peace, and Staff Sergeant Possy (left), during the September 9, 1944 liberation of Moresnet, Belgium –just 7 miles from Clermont.
Photo courtesy of Remember 39-45 Museum.

Bottom photo shows Henri-Chapelle American Cemetery in La Clouse, Belgium. Photo courtesy of author Tom Stein.

Published by

TMS
SERVICES
LLC.

Contact Tom Stein via email: tsteinmd@verizon.net

ISBN: 979-8-9893585-0-2

Printed in the United States of America
First Edition

Cover / interior design, maps, and charts
by Michael Karpovage of
KARPOVAGE CREATIVE, INC.
KarpovageCreative.com

Dedicated to my wife, Janet. Thank you for your patience, understanding, support, and hugs.

And to Courtney and Ryan: I sincerely regret missing milestones and accomplishments due to mobilizations and early book research. I am very proud of both of you.

This is a work of non-fiction. All efforts were made to present accurate information. When possible, primary sources were used. However, as it pertains to the Belgian interviewees, in most cases it is impossible to find a second source to corroborate the details of an individual's story.

The soldier's stories, as far as military action is concerned, were corroborated as much as possible through contemporaneous Company Morning Reports, Battalion Journals, Regimental Reports and After Action Reviews.

The Chart on pages 2 and 3 will assist the reader with perspective of the soldiers and units as they move through history and interact with the civilians. Readers that are unfamiliar with military units, ranks and symbols should refer to this chart as needed to maintain perspective.

CONTENTS

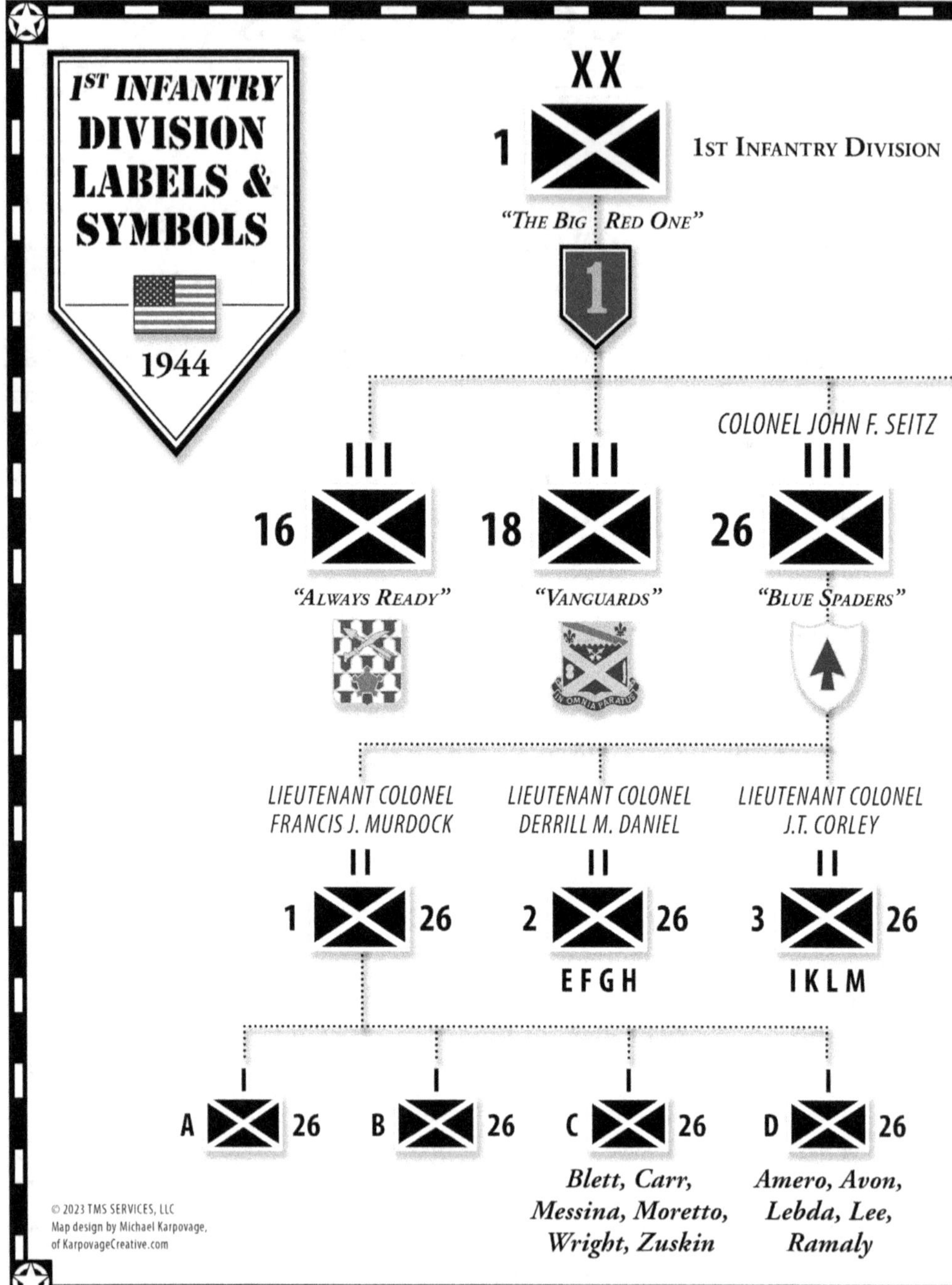

1ST INFANTRY DIVISION LABELS & SYMBOLS
1944
XX
1
1st Infantry Division
"The Big Red One"
1
COLONEL JOHN F. SEITZ
III
16
III
18
III
26
"Always Ready"
"Vanguards"
"Blue Spaders"
LIEUTENANT COLONEL FRANCIS J. MURDOCK
LIEUTENANT COLONEL DERRILL M. DANIEL
LIEUTENANT COLONEL J.T. CORLEY
II
1 26
II
2 26
E F G H
II
3 26
I K L M
I
A 26
I
B 26
I
C 26
Blett, Carr, Messina, Moretto, Wright, Zuskin
I
D 26
Amero, Avon, Lebda, Lee, Ramaly
© 2023 TMS SERVICES, LLC
Map design by Michael Karpovage, of KarpovageCreative.com

3rd Platoon 607th Quartermaster Graves Registration Company

•••

3 607

Mastrangelo

99th Infantry Division
Chesnick, Stanger

1st Engineer Combat Battalion

Clements, Tait

SYMBOL	UNIT	OFFICER RANKS
XXXX	Army	General
XXX	Corps	Lieutenant General
XX	Division	Major General
X	Brigade	Brigader General
III	Regiment	Colonel, Major
II	Batallion	Lieutenant Colonel
I	Company	Captain, 1st Lieutenant
•••	Platoon	2nd Lieutenant

ENLISTED RANKS

Master Sergeant, Sergeant First Class,
Staff Sergeant, Sergeant, Corporal
Private First Class, Private

BELGIUM PROVINCES, CITIES, AND ANNEXATION BORDER
1940 to 1944
NORTH SEA
Brugge
Antwer
OOST VLAANDEREN
Ghent
Scheldt R.
WEST VLAANDEREN
Leie R.
Escault R.
BRUSSELS
Brussels
BELGIUM
Louvai
Lille
HAINAUT
N
E
W
S
Mons
Charlero
Sambre R.
FRANCE
Escault R.
Bavay
Maubeuge
Amiens
Somme R.
Oise R.
Miles
0 10 20 30
0 10 20 30 40
Kilometers
© 2023 TMS SERVICES, LLC
Map design by Michael Karpovage, of KarpovageCreative.com
Oise R.
Aisne R.

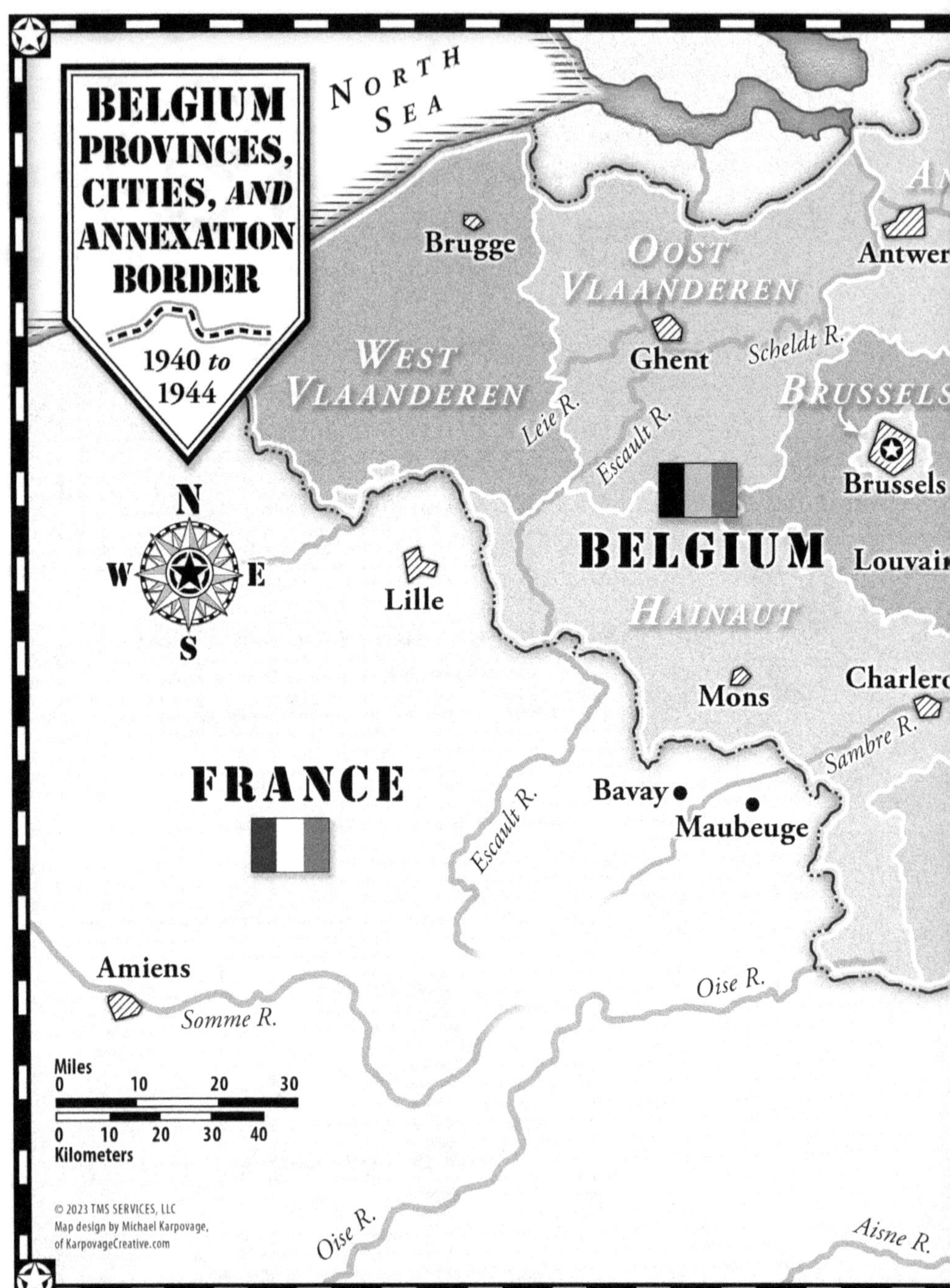

MAP 1

NETHERLANDS
GERMANY
LUXEMBOURG
WEST WALL
Rhine R.
Maas R.
München-Gladbach
Düsseldorf
ERPEN
Scheldt R.
LIMBURG
Meuse R.
Cologne
Hasselt
Leuven
Maastricht
ANNEXATION BORDER
Roer R.
Erft R.
Rhine R.
RABANT
Aachen
Düren
Bonn
Nütheim
Clermont
Liège
Eupen
Verviers
Mosel R.
Huy
Namur
Meuse R.
LIÈGE
Spa
Malmédy
Ourthe R.
NAMUR
St Vith
Dinant
Meuse R.
ANNEXATION BORDER
A R D E N N E S
Our R.
Mosel R.
Bastogne
Bitburg
Bollendorf
LUXEMBOURG
Trier
Arlon
Mosel R.
Meuse R.
Luxembourg

AUTHOR'S NOTE

THIS IS NOT A WAR STORY, ALTHOUGH THERE ARE accounts of war. It is not an historic tome, though there are historic events described. It is the story of how war and history shaped a family's powerful love for each other, for strangers and for a people an ocean away.

It is a story of deprivation, atrocity, fear, hope and gratitude. But, at its core, it is a love story. The story of how war would intertwine a humble Belgian family and a group of American soldiers. A story of hearts touched, relationships made, and unbreakable bonds forged. Bonds that have lasted almost eight decades. It is the story of a farm boy, a lifelong bachelor, who eventually finds his soul mate. And how that union became the catalyst to memorialize those soldiers in an amazing way.

Like my father, I have always had a great interest in World War II, particularly the European Theater. For decades, we swapped our favorite World War II books. Dad was drafted into the Army at 18, but was too young for WWII. And, his unit was on orders to go to Korea when hostilities ended. I spent 38 years in the Army, too young for Vietnam, almost too old for the wars in Iraq and Afghanistan. But I was an Emergency Physician and was able to backfill our Army hospitals in the United States, as well as serve three deployments to Landstuhl Regional Medical Center in Germany. The latter put me in the midst of the battlefields I had read so much about and had always longed to see.

In 2009, to celebrate the 65th Anniversary of D-Day, there was a two-week guided tour of Normandy and the Battle of the Bulge. I was between deployments and wanted Dad to see these beautiful, historic places before it was too late. It would be his first trip out of the United States.

Our tour guide was William (Will) C. C. Cavanagh, a published expert on the Battle of the Bulge. One of our overnight stops was in the city of St. Vith, in the German speaking part of eastern Belgium. Will had scheduled a private dinner for our small group in the Hotel Pip's dining room. He would use dinnertime to review the day's material, answer questions and prepare us for the next day's itinerary.

As part of the evening's festivities, Will had invited a Belgian couple to join us. He introduced them as "The M&Ms," Marcel and Mathilde Schmetz. Both were wearing American Flag accessories, Mathilde a red, white and blue scarf and Marcel a matching tie. Will told us they owned one of the most important World War II museums in Europe. On the tour, we went to several museums, all with the requisite helmets, weapons and uniforms. But this museum, Will said, was something special. He said the "Remember 39-45 Museum" had the requisite war "stuff," but it had so much more.

The Remember 39-45 Museum had stories. Stories of the men who wore those uniforms and used those weapons. The M&Ms knew these men personally, having met most of them as aging veterans when they came to visit their old battlefields. But Marcel also knew some of them when they were young American soldiers. He was 11 years old then, and those men had lived on his family's farm in the weeks before the Battle of the Bulge.

Naturally, all of us in the Hotel Pip's dining room that night wanted to experience that museum. But Will had only teased us. There was no room on the itinerary for a trip to the museum. For me, that would have to wait until my next deployment to Germany in 2012.

After inprocessing at Landstuhl Regional Medical Center in 2012, I began searching for like-minded World War II buffs who might want to explore the battlefields with me. I hit the jackpot when I met an American couple who were experts on the Battle of the Bulge. Mo and Ann Shields showed me places you would never find on any tour. But, more importantly, they turned out to be very good friends with the M&Ms.

My first visit to the Remember 39-45 Museum was chauffeured by Mo, and we actually stayed overnight in Marcel and Mathilde's home. Mo had been there so many times, there was a sign over one of the bedroom doors that proclaimed, "The Mo Shields Room." In fact, it was the same room in which an American soldier had slept after the Battle of the Bulge. The next day Mathilde took me on a private tour of the museum for more than four hours. She was my personal docent. And I heard the stories.

That would be the first of many visits. With each visit I heard more stories of different soldiers. For Mathilde and Marcel, there were no note cards or scripts. Only amazing stories that came straight from the heart. Before long Marcel was sharing personal stories of himself and his family. While the soldiers' stories were remarkable, I came to realize that the story of how these stories came to be, why they are told and why they must continue to be told is even more remarkable. This book is that story.

I began to write down notes about what Marcel and Mathilde had shared with me. I flew to Belgium on several occasions, taking a different friend with me each time so they could experience the museum and the M&Ms. With each trip my notes expanded. I realized even more how truly genuine the M&Ms are in their love for America and its soldiers. It is not just something they turn on when the museum is open. They live it every day even when no Americans are present.

By 2020, I felt I had become close enough to my friends that I could cross the personal-questions line. I knew Marcel was in his late 80s and still did almost all the maintenance on the museum. I was horrified to confirm my suspicion that the legacy plan for the museum was essentially non-existent. There are no family members who could take over operations, let alone articulate the stories. I humbly asked if they would trust me to write their story. And they agreed. My purpose is to bring the M&Ms' story to as many people as possible. My hope is that visibility will somehow generate enough interest to help preserve the Remember 39-45 Museum and its soldiers' memories.

This story is mostly told from Marcel's perspective, up to the 1990s. I was also able to arrange interviews with several Belgians from the same area, most of whom were alive during the war. My pursuit has come too late to interview the soldiers personally. Vito Mastrangelo is the only exception. I was able to speak with him once by phone, and he had graciously agreed to a second session. But Vito passed away shortly before that could happen, just shy of his 99th birthday.

Some of the other soldiers had formally or informally written their memories, and their wives or children have shared them with me. So, much of the information contained here is still from a primary source.

The M&Ms have worked tirelessly to see that these men are never forgotten. This is my small contribution to help them in that mission.

Belgium

Today, Belgium is a small country in northwest Europe with a land mass slightly smaller than the state of Maryland, but with a population twice as large at almost 12 million. Belgium has had a representative democracy with a hereditary monarchy since the early 19th century.

The kingdom—located between Germany and northern France—was uncomfortably positioned for the first half of the 20th century. This was due to the French-German enmity, an international animosity between two countries that began in the 16th century, and became more palpable in 1871. It was then that France lost the Alsace and Lorraine regions to Germany after the Franco-Prussian War.

Throughout that period, Belgium prided itself on its neutrality. Though it had a standing army and had developed defensive strategies, its neutrality prevented Belgium from integrating those strategies with its much larger, and geographically and culturally close, French neighbor. It also excluded the British.

How, then, did this small, neutral country become entangled in two World Wars?

The assassination of Archduke Ferdinand—the heir to the Austro-Hungarian throne—by a Serbian national in June 1914 triggered the chain reaction that began WWI. The Austro-Hungarian government wanted to wipe Serbia from the map, because the Serbs fomented unrest among the Slavic population in the Austro-Hungarian Empire. The assassination gave them the pretext.

The Germans, who had an alliance with the Austro-Hungarians, agreed with this "smaller war." Soon after, with the approval of France, the Russians joined with their fellow Slavs in Serbia. The French-Russian alliance dated back to 1894, its intent to form a two-front foe that could keep Germany in check. Unfortunately, it resulted in Germany declaring war on both Russia and France. The Germans did not want to violate the neutrality of the Netherlands, nor did they want to face the significant defensive line along the French border. This made Belgium the path of least resistance, even though it had been neutral since an 1839 European international treaty.

This avenue of attack was the basis of the Schlieffen Plan, which had been part of Germany's war plans since at least 1905. Its intent was to quickly defeat France, then turn on Russia, effectively avoiding a simultaneous two-front war. On August 2, 1914, the day before Germany declared war on France, the German ambassador in Brussels requested free passage for German troops through Belgium. This, they knew, would make it easier to invade France and get to Paris.

The Schlieffen Plan

Germany gave the Belgians an ultimatum: agree or suffer occupation as an enemy nation. On August 3, the Belgian foreign minister flatly refused the ultimatum. Germany invaded the next day, and Belgium was wholly occupied by November. It remained occupied until November 11, 1918 when an armistice was signed. The last of the German army was gone by November 23.

The Belgian people suffered greatly during those four years of occupation. As the Germans fought their way through Belgium to France, they committed many atrocities. By 1915, only a small strip of Western Belgium was held by the Belgian Army and their French allies. The bulk of occupied Belgium was ruled by a German military governor, but the local administration was kept intact to prevent having to divert German soldiers from the primary war effort.

The closer you were to the fighting line, the more difficult basic existence became. Belgium was a highly industrialized nation before the war, production that was quickly seized by the occupiers. As war dragged on, the Germans resorted to forced labor deportation to maintain output from the German armament factories. Food became scarce. Without food donations from the—then neutral—United States and a few other countries, many more Belgians would have starved or suffered severe malnourishment. This misery shaped the perspective of a generation of young Belgians who, unfortunately, would live to see the Germans again.

In 1933, Adolf Hitler and the Nazi Party came to power, and soon gave indications that German geographic expansion was imminent. Lebensraum, a Nazi doctrine that literally translates to "living space," dictated the amount of territory that would be needed to house all Germanic peoples for a Reich that was to last one thousand years.

In March 1935, Germany reintroduced military conscription. A year later, German troops marched unopposed into the demilitarized Rhineland, an area of western Germany along the Rhine River. Two years later, in what is known as the Anschluss, German troops rolled into Austria—again unopposed—and annexed it as part of Germany. Then, during the Munich Conference in September 1938, Britain and France infamously handed Hitler the Sudetenland, a part of Czechoslovakia primarily populated by ethnic Germans.

By March 1939, the rest of Czechoslovakia ceased to exist, having been divided up between Germany, Hungary (its WWI ally) and Slovakia, a new ally. It was an ominously clear pattern. The original Schlieffen Plan was being updated by Hitler's generals. Soon the world would be introduced to Blitzkrieg, an intense, mechanized military campaign intended to bring about swift victory. On September 1, 1939, German troops invaded Poland, overwhelming them in just over a month. World War II had begun.

At that time, Belgium had a field army of 600,000 in 18 divisions. Only two were motorized cavalry divisions. Only two of the rest were even partially motorized. There were few tanks and essentially no air force. Once again, because of its neutrality, Belgium had no military allies. Hitler and his generals had made the world nervous, but none were more nervous than the Belgian people. They had just watched the much larger and more powerful Poland get overrun nine months earlier. The parallels to 1914 were haunting.

As it still is today, Belgium was made up of provinces. These were made up of communes (municipalities), sometimes consisting of clusters of smaller villages in the rural areas. Most Belgians lived in urban areas, cities such as Brussels, Liege, Antwerp, Bruges, Ghent, Namur, Dinant, Waterloo and Bastogne. A significant part of the country was agricultural. There a minority of the Belgian population earned a critical living.

This story takes place primarily in the rural village of Clermont, in the commune Thimister-Clermont, in the eastern Province of Liege. Less than 16 kilometers (10 mi.) from the German border, Clermont is in a direct line of any attack on the city of Liege. Liege is an industrial and railroad center on the Meuse River, and a strategic military target for any attacker. The young Belgians who had watched German soldiers tear through their towns in 1914 were now middle-aged. Two of them, Henri Schmetz and Maria Grailet, had married shortly after the war and were working hard as dairy farmers, and raising two boys on their farm on the outskirts of Clermont. ✪

SCHMETZ COMMUNITY
OCCUPIED
vs.
ANNEXED
OCT. 1940
Aubel
RUE DE BATTICE
BELGIUM
La Clouse
Berwinne Creek
Froidthier
ELECTRIC TOWER
Henri-Chapelle
Crawhez
QUOIDBACH
SCHMETZ FARM
REMEMBER 39-45 MUSEUM
LES BÉOLLES
La Minerie

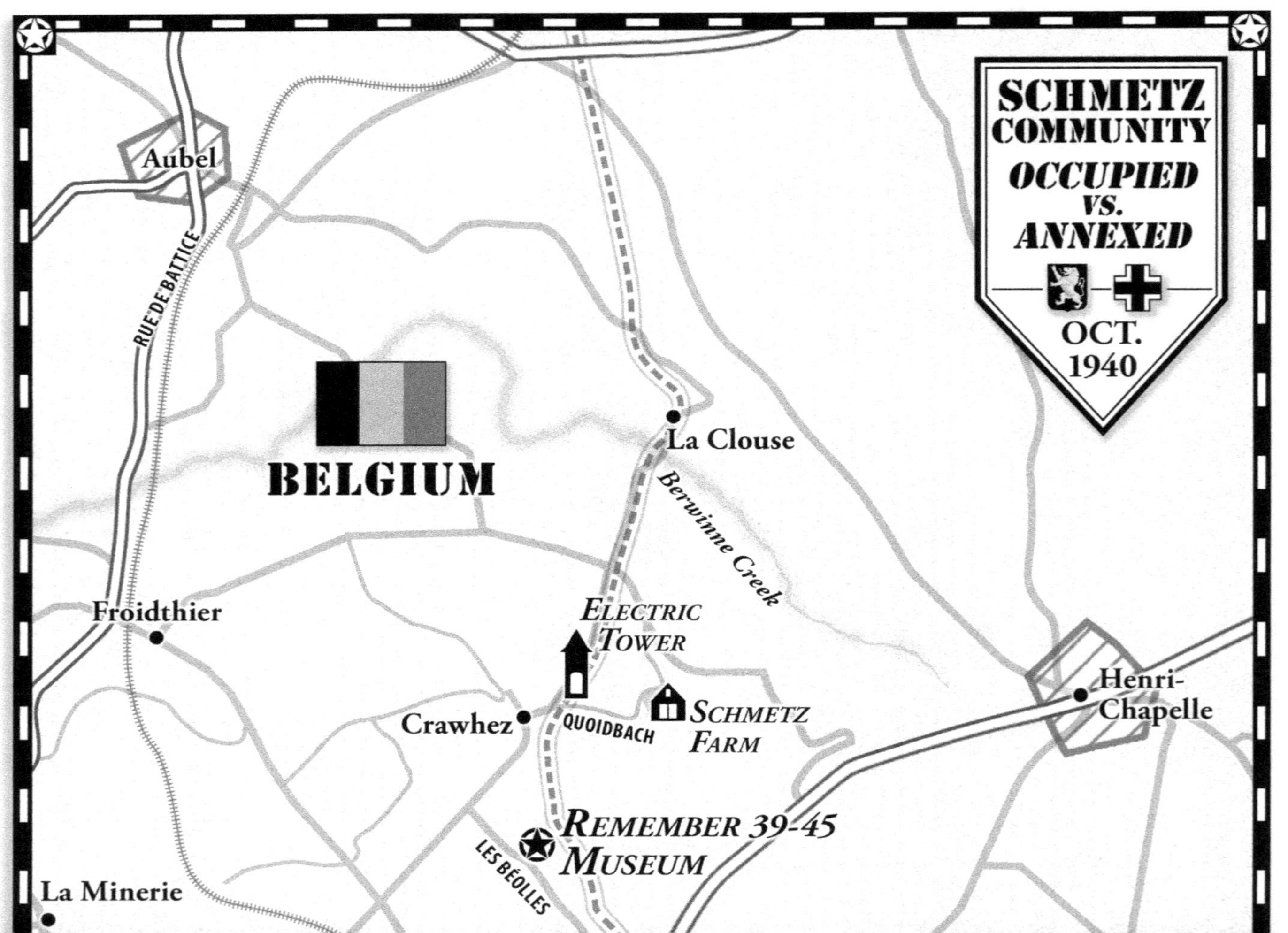

MAP 2

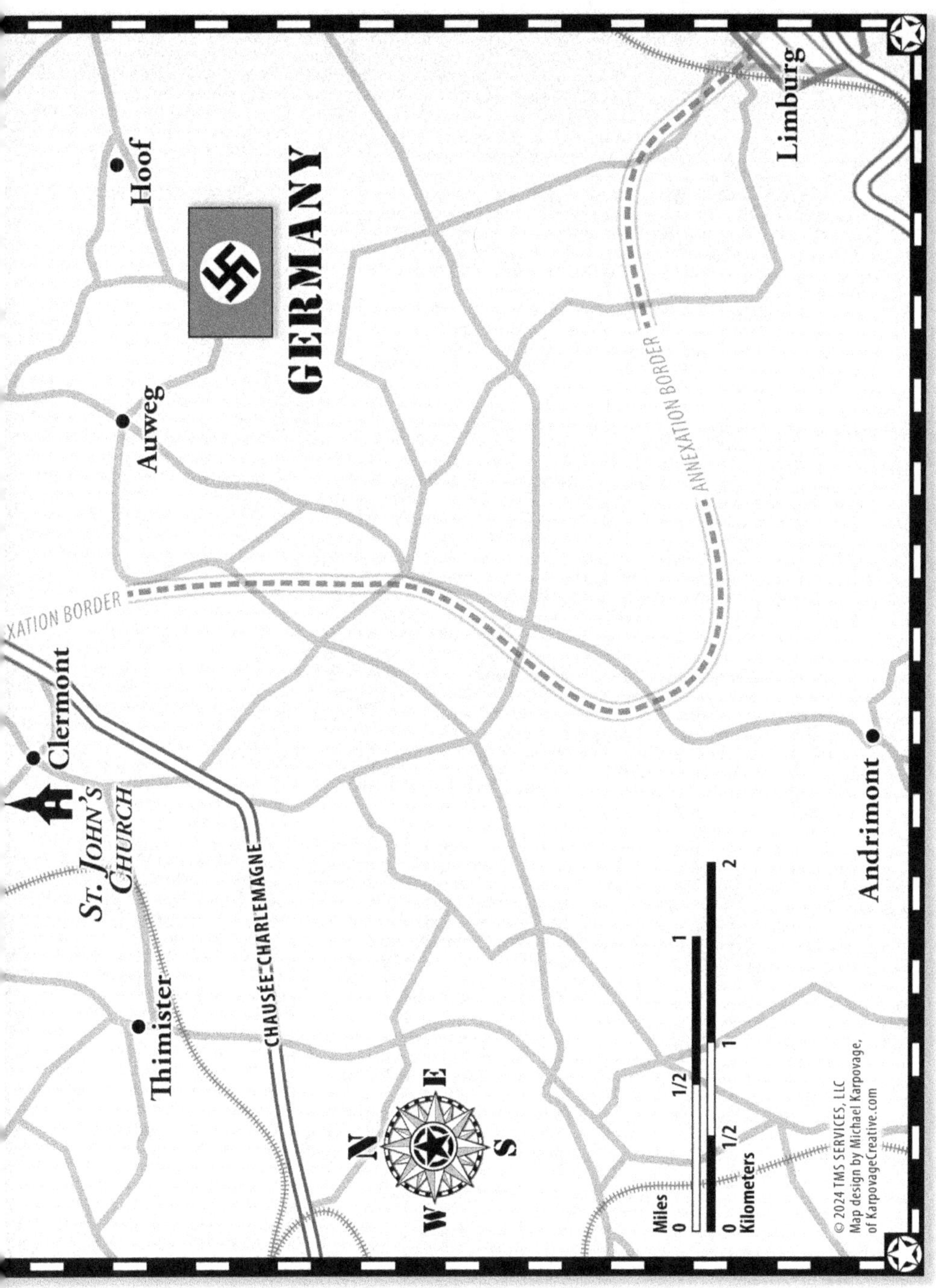
Hoof
GERMANY
Auweg
ANNEXATION BORDER
XATION BORDER
Limburg
Clermont
St. John's Church
Thimister
CHAUSSÉE-CHARLEMAGNE
Andrimont
N
E
S
W
Miles
0
1/2
1
Kilometers
0
1/2
1
2
© 2024 TMS SERVICES, LLC
Map design by Michael Karpovage,
of KarpovageCreative.com

PROLOGUE

MARCEL AND MATHILDE SCHMETZ WERE MARRIED on June 6, 1992, the 48th anniversary of D-Day. Marcel had spent the previous 59 years of his life as an avowed bachelor. For Mathilde, 15 years his junior, it was her second marriage. Together they are the perfect couple.

After their wedding, Mathilde began the task of turning Marcel's house into their home, smoothing some of the rougher masculine edges and adding her own feminine touch. One day, after completing the home makeover, Mathilde decided to surprise Marcel by cleaning and organizing his garage, as well.

Deep in the rear of the dusty garage—past the car, Marcel's tools and other predictable garage paraphernalia—Mathilde discovered a large pile of green duffel bags and a cache of old military-looking items.

For several days, she pondered the intriguing findings. Then, one evening after Marcel had finished with work, Mathilde quietly asked him about the unusual items she had found in the garage. After several long, deep breaths, Marcel told her the most awe-inspiring story she had ever heard.

November, 1943
Belgium-Germany Border

The sky was crystal clear on a chilly November afternoon in 1943 as 10-year-old Marcel Schmetz and his friends walked home from school along Rue Du Bach. As his belly grumbled, Marcel wondered aloud what his mother might be serving for dinner. There was not much to prepare in those days, as the Germans had systematically appropriated most of what the local farms produced to feed the Wehrmacht (the German Armed Forces). Marcel's parents' farm was no different. In fact, Gerard Krafft, the local man the Nazi's had entrusted to implement their draconian policies, was literally their next-door neighbor.

The boys were approaching the small bridge over Berwinne Creek, from where, after crossing, Marcel would cut through the fertile fields to his house. Suddenly, they all saw it: two German border guards standing over the body of a young woman in the creek. She was motionless, face down in the water. The water just downstream was an odd pinkish color.

The guards' weapons were not shouldered, as they usually were, and the young lads knew immediately what had happened. Without prior authorization, it was forbidden for locals to cross the arbitrary border line Hitler had drawn through Marcel's village. It could also be considered a capital crime to smuggle food across that line, depending on the whims of the border guards. Just meters from the bridge to the west was occupied Belgium. To the east was considered Germany. And all along that line there were checkpoints with armed guards and, sometimes, vicious Shepherd dogs.

The woman had attempted to sneak across the fields with what those guards considered contraband. Marcel and his friends would never forget the haunting image of that offending loaf of bread, bobbing in the creek. ✪

1

TWO INVASIONS

BY EVERY MEASUREMENT, AUGUST OF 1914 WAS considered a scorcher. The morning air was already stifling as 14-year-old Henri Schmetz approached the cows for his morning milking routine. The cows were Henri's first priority every morning during summer vacation. He wished he could defer the task and sleep in, but he knew the cows' schedule superseded his own comfort, which, frankly, was low on the dairy farm's priority list.

Following the milking, he would also have to tend to the fields and stables in even more oppressive heat. One full milk can, with a little straw floating on the top, would go to the laiterie, where Henri's mother would process the milk, make fresh butter and use the separated buttermilk to make her delicious waffles.

As the sun rose, Henri heard not-too-distant booms followed by the crackle of firecrackers. Though just a teenager, he was fully aware of the German saber-rattling since the June assassination of Archduke Ferdinand in Sarajevo. But Belgium was neutral and—though the King had recently mobilized the army as a precaution—Henri assumed the Germans would honor this well-known policy. He was very wrong. The beautiful village of Clermont was directly in the path of the German Army on its way to Liege, the industrial center of Belgium, and a very important strategic target.

Henri was born in Clermont in 1900 on the family's dairy farm in the plush, rolling hills of eastern Belgium. Though the farm was just a few kilometers west of the thickly-forested, traditionally German-speaking, Ardennes region, the Schmetz family—like most Belgians living south of Brussels—spoke French.

Dairy farming was a hard life, but it was enjoyable in this idyllic setting.

From his father, Henri learned the ways of the dairy farmer: twice-a-day milking, livestock feeding and tending of fields and equipment seven days a week, a lifestyle he would eventually teach his own sons. That summer, Henri was too young for military service.

But for the next four, grueling years, he would still experience the fear and deprivation of enemy occupation. If there was a small consolation to being conquered and occupied early, it was that the farm was far behind the brutal trench warfare that would devastate parts of Belgium and France, dragging on for four long years.

When he turned 18, Henri was conscripted into the Belgian Army, just as the Great War was ending. He was deeply thankful the shooting was over. His 18-month compulsory service was spent as part of an occupying force in Germany's industrial Ruhr Valley. The British, French and Belgians split the Ruhr into zones to keep it—and the rest of Germany—demilitarized. Europe could not afford another "world war," having already lost one full generation of young men.

Though he was now the occupier, Henri held no animosity toward the German people he encountered during his tour. Even though they were not friendly, the locals were not belligerent either. Their homes stood intact, unlike so many in Belgium and France. But the vanquished country was financially spent, and food and staples were hard to come by.

Henri keenly observed how the Germans thought and behaved, which was very different than his family and neighbors back home. They seemed more regimented, more formal. He did not know if this was a cultural difference or if the loss of a major war had changed them. He also did not realize at the time just how well this insight would serve him in the future.

A New Schmetz Family Is Formed

As the end of his enlistment neared, Henri looked forward to going home, continuing his family's work on the dairy farm and perhaps starting a family of his own. At that time, rural Belgian communities were just like most rural communities around the world. The geographic universe of potential spouses was pretty much your own community, and those immediately surrounding it.

Nevertheless, shortly after returning home, Henri met a young woman—Maria Grailet—one year his senior. Maria's father was also a farmer in Clermont, and he expected his daughter to know and perform the routine chores of the farm. In her spare time, Maria also took sewing lessons from one of the local seamstresses. In 1925, the young couple was married, and began life on their own dairy farm.

Maria and Henri Schmetz, circa 1924.

As a teenager, Maria had experienced firsthand the horrors of war. In 1914, a close friend of hers watched from her window as the Germans passed through the streets of Clermont. Suddenly, and for no reason, the invaders began shooting randomly into houses. Maria's friend was shot and killed.

Nearby, a neighbor was shot while cutting firewood in his backyard. German soldiers often confiscated alcohol from the farms, drinking until drunk. This was often a contributing factor in some of the atrocities. When discipline was lax, some soldiers ran amuck. They randomly burned homes in the neighboring villages of Battice and Herve, gunning down the occupants as they fled the flames.

War crimes, otherwise known as murders in times of peace, were committed throughout Belgium, particularly in the early days of WWI. On the order of a German Corps commander, 674 civilians were executed in the town of Dinant.

The city of Louvain, including its famous Louvain University library (known then as the Université catholique de Louvain), was burned to the ground. The burning of the library, a cultural treasure established in 1425 that contained books from the 13th century, became the focus of profound political outrage. Far worse, many of Louvain's citizens, including women and children, were executed.

By war's end, more than 5,500 Belgian civilians had been murdered. The memory of those tragic events was imprinted on the Belgian psyche and would color their reactions less than 20 years later. Maria never forgot what the Germans could, and would, do to civilians.

Henri's and Maria's farm was located on Rue Du Quoidbach, a winding country road that still runs between Clermont and La Clouse. A large brick and stone house dominated the front of the property. A solid wall divided the building in half; Henri's grandmother lived on the other side. A large barn angled perpendicular from the house toward Rue Du Quoidbach on Henri's side of the building.

The property was owned by a Liege businessman who rented it to Henri and Maria. In 1939, Henri's grandmother died. Gerard Krafft and his family, originally from Germany, moved in as the second tenant, and the Schmetz's new neighbors. Henri and Gerard became good friends and worked adjoining fields.

The Schmetz Farm, circa 1930s.

The "summer kitchen" is on the far left across from the front door. The dividing wall that separated the Schmetz family from the Krafft family is just to the left of the fifth window on the upper floor.

The Schmetz Farm today, owned by Charles Putters.

The Schmetzes Start A Family

In 1926, Henri and Maria's first child—Henri, Jr.—was born into the family dairy farming tradition. Like his father, young Henri would be 14 years old when another world war erupted. Seven years after Henri, Jr. was born, the Schmetzes welcomed their second son, Marcel. It was the same year as the birth of Hitler's so-called Thousand Year Reich: 1933.

Young Marcel grew up happily with his brother and other nearby children. Three neighbors had boys with whom Marcel and Henri, Jr. would play and go to school. All the boys lived and worked on family farms, learning the skills needed to one day become successful farmers themselves.

In winter, that meant cleaning the stables and barns and making necessary repairs. Summers were spent in the fields. Year-round, they milked the cows three times a day, as opposed to today's modern schedule of twice per day. For fun, the boys played football (soccer) in the fields or streets. On rainy days they drew pictures and played board games.

The Schmetz boys went to school in the nearby village center of Clermont. They walked the country paths and roads to and from school, about a kilometer each way. They would pass the beautiful town square, where St. John's Church towered over the town, and the city hall arched over the main street. Between the two was the brasserie, the town's simple, informal restaurant and de facto "neighborhood pub."

Henri, Jr., Marcel and friend Rene.

The Rumblings Of War...Again

When Marcel was old enough to begin school, Nazi Germany was casting an ominous shadow over Belgium. And his parents were still acutely aware of the route the German Army had taken to reach France in 1914. To mitigate any new threat after World War I, the Belgian government planned a network of six major fort systems along the border facing Germany. The fort system was designed to tie in with older forts that were built before World War I, effectively encircling Liege. By 1940, only four of them were complete.

Fort Battice, one of two larger forts—along with Fort Eben-Emael, located to the north near Maastricht—was located a mere three kilometers (1.9 mi.) from Clermont. Fort Battice boasted two 120mm guns and six 75mm guns mounted in rotating turrets that could be lowered into the fort for reloading. Although Clermont was located between the fort and Germany, the townspeople felt safe knowing the big guns could fire over them and hit invaders long before they could reach the town.

Similarly, France sought to prevent another German invasion by developing an even more robust Maginot Line in the 1930s. This was an interlinked line of heavily armed, armored bunkers, gun emplacements and machine gun nests designed to exact a heavy toll on any invader. Or, hopefully, discourage the Germans from trying in the first place. Unfortunately, it did not extend along the French border with Belgium, as neutrality prevented the Belgians from tying into the Maginot terminus. For Hitler, this made Belgium an even more attractive gateway to France. Though the where and when were unclear, the Belgians were anticipating another German invasion.

Henri, Sr. and Maria, like many Belgians, thought the Belgian government was not at the top of its diplomatic game. Why would they think that way? One reason was that the Belgian government had hired German workers to assist in the construction of the forts. So, the German Wehrmacht not only knew the precise fort locations, but they also knew the inner workings of their defensive fortifications.

The Germans Invade

Claiming he was merely responding to provocations by the Poles, Hitler invaded Poland on September 1, 1939. To Henri and Maria, it seemed far away, and they hoped Poland, along with the annexation of Austria and the Sudetenland, would be enough to appease Hitler. After all, the Belgians had made no provocations toward Germany. But a little more than eight months later, at 0500, May 10, 1940, the Germans invaded their tiny Belgian neighbor. Again, as a through-way to France. This time the Manstein Plan was used, approximating the reverse of the WWI Schlieffen Plan. The common denominator was Belgium as a door to France.

That morning, Henri and Maria were aware of the news by the time they had to wake the boys for school. Ordinarily Maria would wake the boys, as Henri would already be tending to his morning farm routine. That was why Marcel was confused when both parents were at his bedside, tears in their eyes. In a scene vividly etched in Marcel's memory, Henri kissed each of his boys goodbye, and headed off to war. Marcel and Henri, Jr. would not be going to school that day.

Henri, Sr. and Maria, post-war.

Henri, Sr. Joins The Fight

Henri, Sr. was 40 years old, too old for the infantry. But he could drive a truck. At that time, not many people could drive a vehicle, and even fewer owned one. The Belgian Army needed every adult male to help form any significant defense against the German onslaught. And they needed every available truck to get that defense to the right place at the right time. Previously Henri had been instructed where to report for duty in the event

SCHLIEFFEN INVASION PLAN OF BELGIUM & FRANCE
AUG. 1, 1914
N
W
E
S
London
BRITAIN
Dover
Calais
Dunkirk
ENGLISH CHANNEL
Somme R.
Rouen
NORMANDY
Seine R.
Oise R.
Paris
Chartres
© 2023 TMS SERVICES, LLC
Map design by Michael Karpovage, of KarpovageCreative.com

MAP 3

Amsterdam
NORTH SEA
NETHERLANDS
Rhine R.
Meuse R.
Antwerp
BELGIUM
Scheldt R.
GERMANY
Brussels
Eupen
Liège
Verviers
Mons
Namur
ARDENNES
Maubeuge
Meuse R.
Our R.
Mosel R.
Bastogne
LUXEMBOURG
Soissons
Aisne R.
Reims
Verdun
Meuse R.
Marne R.
FRANCE
Miles
0 20 40 60
0 20 40 60 80
Kilometers

MANSTEIN INVASION PLAN OF BELGIUM & FRANCE
MAY 10, 1940
N
W
E
S
London
BRITAIN
Dover
Calais
Dunkirl
ENGLISH CHANNEL
Somme R.
Rouen
NORMANDY
Seine R.
Oise R.
Paris
Chartres
© 2023 TMS SERVICES, LLC
Map design by Michael Karpovage, of KarpovageCreative.com

MAP 4

Amsterdam
NORTH SEA
NETHERLANDS
Rhine R.
Meuse R.
Antwerp
BELGIUM
Scheldt R.
GERMANY
Brussels
Eupen
Liège
Mons
Verviers
Namur
ARDENNES
Meuse R.
Maubeuge
Our R.
Mosel R.
Bastogne
LUXEMBOURG
Soissons
Aisne R.
Reims
Verdun
Meuse R.
Marne R.
FRANCE
Miles
0 20 40 60
0 20 40 60 80
Kilometers

of an invasion. There he was assigned to drive a re-purposed, red truck the Army had not had time to repaint. Henri could not help feeling like he was driving a rolling target.

Henri's job was to pick up ammunition, food and other supplies and deliver it to the troops. He was not directly involved in the front-line fighting, but any vehicle on the road was fair game for Luftwaffe pilots who had complete control of the air. And German artillery could easily reach the forward supply depots. As May marched on, the Belgians fought bravely, but had to continually fall back to new defensive lines. Their final redoubt would be the port of Antwerp, where their backs would be against the water.

"I Would Rather Die In My Home Than Die In The Streets"

Remembering the suffering of 1914 and the behavior of the German Army, 70% of the local Belgian population evacuated toward France. Maria chose to stay in her home with her sons. She had always said she would rather die in her home than die in the streets, running from the Germans. Maria's intuition proved telling, as many Belgians, including women and children, were killed on the open roads by advancing Germans who sometimes strafed, bombed, and shelled indiscriminately. Many others died of illness as physicians, medications and medical supplies were first sequestered for the Belgian Army, then ultimately confiscated by the Germans.

After only 18 days of fighting, Belgium's King Leopold III capitulated. He knew any further fighting would only kill many more of his countrymen. That was May 28.

Henri's red truck was lucky enough to remain unscathed. He fell back from defensive line to defensive line until he could fall back no more. Ultimately, he was captured near Antwerp and disappeared to the east with hordes of other Belgian soldiers. Once again, Henri would observe how his German captors thought and operated. This understanding would come to serve him and his family well in the not-too-distant future. ✪

2

OCCUPATION AND ANNEXATION

SHORTLY AFTER THE BELGIAN CAPITULATION ON May 28, 1940, the Germans expanded their western border by drawing a north-south line through Belgium. Prior to the end of World War I, the eastern part of the annexed territory had been part of Prussia and the German Empire. So, the Germans justified the annexation by claiming they were merely taking back the land the 1919 Treaty of Versailles had ripped from them.

The eastern third of the Clermont commune, which was attached to the neighboring commune of Henri-Chappelle, was wholly annexed. This put the western two-thirds, including the Schmetz's farm, in German-occupied territory. Henri, Sr. and Maria clearly remembered how their parents had experienced German occupation for more than four years during World War I. Back then, the Kaiser had ransacked Belgium's natural resources, took over and dismantled their considerable manufacturing capacity and took much of it back to Germany.

Naturally, all were fearful of what was to come. The brief fighting in Belgium had damaged considerable infrastructure. Many fields were torn up by vehicles and bombs, ruining that season's crops for many. Though the Schmetz's land was relatively unscathed, Henri was a prisoner of war and would not be there to do the work. It was up to Maria and Henri, Jr. to carry on. Little Marcel was expected to pull his weight, too.

The Schmetz Farm Becomes German Soil

The summer of 1940 came and went. The invading soldiers moved on to bigger conquests, and both the French and British armies were defeated in France. Any hope for freedom had vanished. In October, Hitler fine-tuned his land-grab in Belgium. A second "new" border added 10 more villages, including the remainder of Clermont, that had previously only been considered occupied. So, on one frigid day in October, the Schmetz's farm became German soil, while their neighbors' farms on the other side of Rue Du Crawhez still remained occupied enemy territory. Just like that, the Schmetz family was German.

The Germans constructed physical barriers across each street that crossed the new line, establishing "check points" for anyone crossing from Belgium to Germany, or vice versa. The combat soldiers were gone but were replaced by uniformed border guards. These were soldiers the Germans no longer considered fit for fighting. But they were considered fit enough to be armed, and they reported to the local Nazi authorities. This created a perplexing new challenge. Because the border guards ranged from rabid Nazis to locals sometimes sympathetic to their occupiers, the burning question for the locals became: Could a uniformed local be trusted? After all, he was a collaborator.

This drawing by young Marcel shows the annexation border crossing and electric tower located in Crawhez at the fork in road with Quoidbach. It was less than 1/4 mile West of their farm. Photo on right shows how the tower looks today.

A Prisoner Of War

From Antwerp, Henri, Sr. was shuffled back through German lines from one prisoner-of-war camp to another. Ultimately, he wound up in a camp located in what had been Poland in 1939 but was now part of Germany. By then, the Wehrmacht had already mobilized all able-bodied men, including many farmers and farm workers. Military prisoners, foreign civilians and concentration camp prisoners were used as forced labor on German farms and in factories, mines and other war-supporting industries.

Henri was sent to several nearby farms. His experiences varied, depending upon the "Bauern Frau," the woman of the farm. As a general rule, this was the wife of a farmer who had been mobilized. Some were ardent Nazis, who would make life miserable for the laborers. Food and drink might be restricted, compared even to what had been served in the camps. In Henri's experiences, most were more considerate, understanding their own loved ones might one day be prisoners and hoping they would be given similar consideration. Henri was never treated too harshly and was able to maintain his strength and positive outlook. Then, months into his captivity, fate smiled upon Henri.

Hitler's new October annexation boundary had officially made Henri...a German! And, as the wheels of bureaucracy slowly turned, he was eventually set free and returned home in January 1941. This was, however, far from a compassionate move by the Germans. They shrewdly anticipated he would work harder and more efficiently on his own farm, providing even greater benefit to his former captors.

Becoming "German"

Annexed or occupied, those on either side of the new border were not treated well. But Henri Schmetz would argue that those considered occupied actually had it better. After all, they were allowed to continue to speak French, and students could learn in their native language.

Conversely, those on the annexed side were strictly controlled. The Germans knew these Belgians did not want to be German, and therefore could not be trusted. In the annexed part of Belgium, teachers and priests were removed and replaced, mostly by Germans, or Nazi-sympathizing Belgians. Young Marcel's area of Clermont was attached to the Catholic parish in the village of La Clouse. Father Corman, the parish priest serving

La Clouse, was Belgian, but a Nazi sympathizer.

In 1940 and 1941, the parents of children of First Communion age quietly asked Father Corman if catechism lessons could be taught in French. He flatly refused. All persons were expected to speak German, use German books in school and listen only to German broadcasts. At least Father Corman did not turn in the parents for violating German law. Undeterred, some of the parents secretly took their children to the Chapel of Gensterbloem, where Father Laduron, at great risk to himself, was willing to give lessons in French.

Annexed citizens were also steadfastly prohibited from locking their doors, and the Germans vigorously enforced their right to inspect a private home at any time, day or night with no knock at the door. A locked door would result in a broken door, and a harsh punishment. In addition, most of the milk, milk products and crops produced on the annexed farms were confiscated for use by the German Army, reallocated to "real" Germans. The Belgians were paid pennies on the dollar for what was taken and were forbidden to sell what little was left.

Marcel Goes To First Grade

The Belgian school year ran from September to June. Marcel began first grade in September 1939 at the age of six. In the beginning, he walked with his brother to school in the Clermont village center. But when his father was recalled to active duty in the Belgian Army, Henri, Jr. had to stay home to help his mother with the farm. When the Germans took over Clermont, the school building was immediately commandeered, and school was shut down. Marcel's teacher, Mr. Bemelmans, along with many others, was dismissed from his job. They were given but a few hours to pack up everything and leave.

The portraits of beloved King Leopold III and Queen Astrid that adorned every classroom were taken down and replaced by Hitler's visage. The school remained closed for three weeks, until after the capitulation. With Clermont and Quoidbach now in occupied Belgium, Marcel was able to return to school and finish the year, albeit with only half-day schedules. In September, he began the second grade. But in October, Marcel and the other "new" German students were reassigned to a "new" German school. Marcel and the children who lived in Quoidbach were required to attend a school in La Clouse, the same town where they were forced to go to church.

Each classroom had a Nazi flag with a large swastika in the center. The morning prayer was replaced by the Hitler salute. And the children went to school without understanding a single word of German.

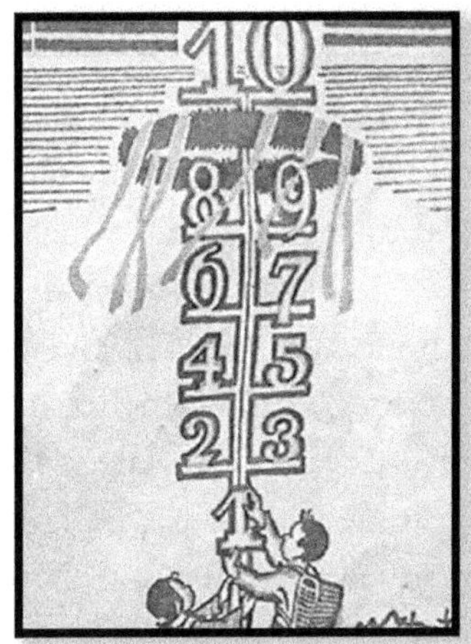

Cover and title page of Marcel's German math book, circa 1940.

Herr Reynold Kreuz, a 30-year-old German from Cologne, became their new teacher. Immediately, he was despised by his students. Kreuz was a card-carrying Nazi who ran his classroom like a military boot camp. Fortunately for the students, to prove that he was a good Nazi, Kreuz volunteered to join the German Army at the Russian front. This gave the children of Clermont a respite from their cruel indoctrination. The hated Kreuz was replaced by Fraulein Hilda Stern, from Bleifeld, Germany, who proved to be significantly more lenient. Even now, Marcel has surprisingly fond memories of his two years with Fraulein Stern.

Obviously, these were not ideal educational circumstances. To further complicate matters, although the opportunity to learn the German language could have been valuable to the Belgian students, some parents who had experienced German brutality during World War I took a passive-aggressive form of resistance. They instructed their children not to learn German, as it was the language of the enemy.

Alice Didden's Story

In annexed Belgium, a few kilometers could make a big difference in how you were treated. Alice Didden, the youngest of seven children, was eight years old when the Germans invaded. Her parents' farm was located in the annexed village of Baelen-Hoof, six kilometers (3.5 mi.) east of Fort Battice, and only two-and-a-half kilometers (1.5 mi.) southeast of the Schmetz Farm. Even though Alice's school was in Elsaute, a town 50 meters (55 yds.) inside occupied Belgium, she was allowed to continue her education there for two years, until a new German school could be built in the annexed village of Wooz. Alice's school was Catholic and retained the customary religious décor, such as crucifixes and statues. But when her new school opened, only pictures of Hitler adorned the walls. On Alice's first day at the new school in the village of Wooz, a single teacher was assigned to 180 children. Fortunately, his fiancé accompanied him, helping to preoccupy those students who were not actively interested in learning.

At first, Alice and her classmates did not realize their teacher was anti-Nazi. For two days, it was all, *"Heil, Hitler!"* But as time went on, there was a gradual change. One morning, the teacher came into the classroom to find Hitler's picture covered in black. *"He was trying really hard not to laugh,"* Alice recalled. No punishment was administered, and a stern, *"Do not do that again,"* was all he could muster. But the door had been opened slightly for a group of rambunctious boys. A few days later—and rather inexplicably—the school's chimney became plugged. The entire building was engulfed in smoke, and the students had to be sent home for several days while their classroom was thoroughly cleaned.

A Farm On The Front Line

In the event of hostilities, the Belgian Army planned to commandeer the Diddens' telephone to assist in directing defensive fire from Fort Battice. The Didden farm was adjacent to a main east-west road and Alice's parents—remembering the carnage they had witnessed when the Germans invaded in 1914—desperately wanted to evacuate if the anticipated war began. However, the logistics of moving nine people and leaving behind their livestock, and livelihood, forced them to reconsider. Alice's father struck a deal with a neighboring farmer friend to move in with them. The neighbor's buildings were far enough back, on a private lane away from the main road, that they would be relatively safe, yet still be able to maintain their farm.

Early the morning of May 10, 1940, Alice was milking a cow when a German soldier approached her. He asked for some of the milk, and the terrified Alice gave it to him. The soldier reassured her that he would cause no harm to her or her family. Later that day, the soldier returned to the farm to warn the family to move out, as fighting was expected to course through the area. Soon, defensive fire from Fort Battice roared overhead. The next morning, Alice's father found two of his cows and five of his neighbor's cows dead. He also found unexploded shells burrowed into the field. The farmers dug a large hole to bury the cows. Then carefully threw the shells in with them, knowing they were fortunate not to have been blown to bits.

More Passive-Aggressive Resistance

Alice's 14-year-old brother, Roger, was ordered to attend Hitler Youth training on the weekends. Her father went to the Kommandantür to ask that Roger be excused. His two oldest sons had been captured earlier, during the short fight with the Belgian Army, and were already imprisoned in Germany. He needed the young boy to help with the farm. *"Your son will work for the Reich,"* the Kommandantür said, laughing.

Mr. Didden fought back with a dogged level of passive-aggression. Because they were annexed, just like the Schmetz family, they were subject to many draconian measures. Farmers were expected to produce crops and goods for the Wehrmacht and the Reich. Provision formulas for every commodity were mandated to ensure the Wehrmacht received its "fair share." The formulas, of course, were anything but fair. For example, if you produced 100 liters of milk each week, you could be expected to turn 90 over to the Germans. Similar allotments were dictated for butter, potatoes, eggs, beef, pork...everything.

At that time, the Diddens were feeding 10 mouths, counting their hosts. Alice recalls that her father—at great risk—would only give half the farm yield to the Germans, far less than their required allotment. The family "log," however, would show they gave far more. By law, Belgian families were not permitted to have fresh butter. Mr. Didden—as was required by law—turned over his butter-churn bowl to the authorities. But he had kept a spare, and the Diddens were able to enjoy fresh butter, while carefully concealing their secret from the Bauernführer, the Farm Controller.

After 1940, Alice did not see German troops again until late 1943. Then, soldiers came to inspect their fields and began digging defensive trenches for a future fallback position. At the same time, Allied bombing became much more frequent. The outskirts of Aachen were only a few kilometers to the northeast. The air attack formations were so large and so dense that Alice could actually see bombs falling on Aachen. Routinely, air raid sirens would cause her school to be evacuated, a prudent precaution, given that errant bombs had fallen on the Montzen train station, just north of Hoof, killing many civilians.

The Controllers

Residents of Clermont had to accept and adjust to people Marcel called, "The Controllers." These were either local Belgian citizens (now considered Germans) or Nazis brought in and appointed to manage the local and regional governments. Everything was under strict German control. The provisions assessment formulas and other rules were constantly updated and distributed to the families by their local Bauernführer. As the war dragged on, these rules became more oppressive and more challenging with which to comply. The Nazis preferred their "citizens" malnourished; it made it that much harder for them to resist or rebel.

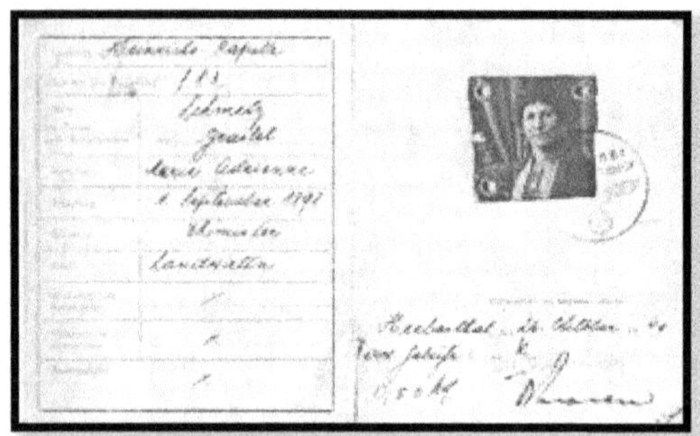

Maria's "New" German ID papers.

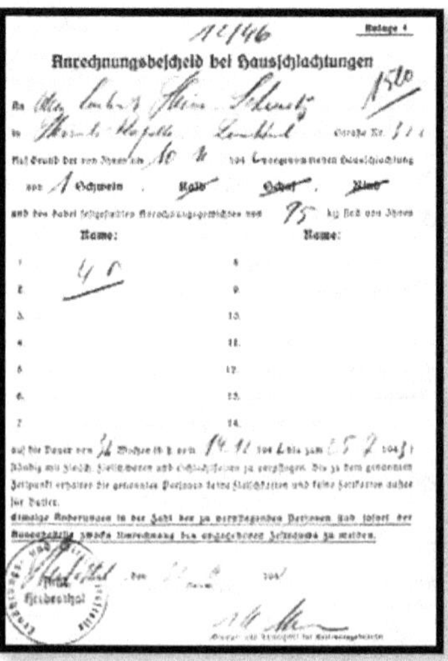

Anlage 4

Anrechnungsbescheid bei Hausschlachtungen

1 Schwein

Name: Name:

Documentation of a pig slaughtered by Henri. (The document certifies that "Henri Schmetz, having slaughtered a 95-kg pig, is deprived of meat supply stamps for 32 weeks' meat for a household of four people. This period runs from 14.12.1942 until 25.07.1943".)

GERARD KRAFFT

Germany had been at war since September 1939. After losing the Battle of Britain in 1940, its cities had been subjected to almost around-the-clock bombardment, first by the British and later the Americans in 1942. So, in an ironic twist of fate, the German citizens—especially those in urban areas—lacked for food as well.

Like many of the structures in the area, the large brick and stone farmhouse occupied by Henri, Sr., Maria and their sons was hundreds of years old. The long building was divided in half by a solid wall, enabling residence by two families with relative privacy. The Krafft family lived on the other side of that wall. Gerard Krafft was German-born, one of four brothers. The oldest, Leonard Krafft, became Mayor of Membach, Belgium, but was recalled by the Reich and transferred to the Eastern Front to fight the Russians. Gerard and his wife Catherine had also lived in Membach, but moved to Clermont in May 1939. Their son, Hubert, was five years younger than Marcel.

Marcel Schmetz and Hubert Krafft, 2023.

When the Bauernführer in Henri-Chappelle decided he no longer wanted to deal directly with the Nazi regime, he retired. So, the Germans turned to Gerard Krafft. They believed him to be more trustworthy than any native Belgian, and appointed him the new Bauernführer. For Gerard, it meant he would not be drafted and sent to the front. The Nazis believed "real" Germans could be trusted to keep the annexed Belgians under tighter control. As it turned out, Gerard was sensitive to the plight of his friends and neighbors. But he was very cautious to never appear less than 100% loyal to the Germans. Even in the toughest times, Gerard and Henri, Sr. remained good friends and good neighbors. They would even go on to conspire against the Nazis.

In 1933, Henri bought a car, an uncommon luxury for that time in rural Belgium. When the Germans stormed through in 1940, they confiscated all vehicles for use by the army. Henri hid his car under hay in the back of his barn, and Gerard never reported his neighbor to the authorities. For his considerate application and "diligent" enforcement of Nazi rules, Gerard was viewed by his neighbors as one of the "good" German controllers. Thanks to Gerard's "blindness," Henri's car remained safely in the barn until the end of the war. For the Nazis, the vehicle-grab turned out to be a rather short-term benefit. Especially once they realized that finding replacement parts for so many different types of vehicles was a logistical nightmare.

Brutal Rules And Regulations

The rules and regulations imposed upon the Belgians were strictly enforced. Violations were punished swiftly and—at times—brutally. One of the crimes the Nazis considered particularly heinous was hoarding food for personal or family use. In 1942, police caught Eugene Kevers in the act of making a half-pound of butter for his family. He was sentenced to 18 months in an Aachen prison, where he performed forced labor in a quarry for the first six weeks. When a nearby Aachen farm owner and his sons were drafted into the Wehrmacht, Eugene was sent to their farm to continue his forced labor.

Six months into his sentence he sustained a significant foot injury that prevented him from meeting his expected quotas. The German foreman banished him from the farm, his only lodging. Hobbled and with no money, Kevers limped back home, 30 kilometers (18.6 mi.) away. In the town of Montzen, still six kilometers (3.7 mi.) from his home, Kevers was spotted by a neighbor, Mr. Charbon, who transported him the rest of the way home in a horse-pulled cart.

Kevers paid a considerable price for his petty infraction. But he was luckier than Guillaume Renard. Renard lived in the Clermont village center and ran a café. One day, police stopped him with a sack of potatoes on his back. He was planning to share them with neighbors who did not have enough to eat. For his indiscretion, Renard was marched to the edge of a nearby hillside, overlooking a road below. He was summarily executed, falling to the pavement below, where his body was left as a reminder for all to see.

Guillaume Renard outside his café.

A roadside memorial at the site where his body laid.

Annexed, But Compassionate

After October 1940, it was illegal to cross from the annexed side of town to the occupied side unless you had written authorization from the authorities. A request for permission was submitted to the Kommandantür and, if approved, you would be given papers with a signature and stamp of approval. These papers were very difficult to fake or copy. With permission, you could cross the "border" for a wedding or funeral, but you were mandated to return before nightfall.

Prisoners and forced laborers were quite another matter. They would escape from prison or labor camps in and around Aachen, then head west. Some were Belgian, but many were French, Dutch, Luxembourgish and even Russian. Henri Schmetz and Gerard Krafft shared the same compassion for these desperate people and conspired to help them. When escapees would show up on the doorstep of the Quoidbach home, the two men would hide them in the haylofts.

At an opportune time, Gerard would invite the border guards to his home for coffee. While Gerard entertained the guards, Catherine Krafft would alert Henri when and where the border was unguarded. Then, at dusk, Henri would help the escapees cross the border. Had they been discovered their scheme would have resulted in a death sentence for all.

One Christmas Eve 1940, the Krafft family was singing German Christmas songs. A border guard posted nearby could hear the merriment. He knocked on the door and asked if he could come in and join them. It was not just the cold that drew him in, but the melancholia he was experiencing. He told the Kraffts that his father and three brothers had been killed in the war, and his mother would be alone for Christmas.

That winter the locals were also notified they would be responsible for snow clearing, anywhere the Kommandantür's bureaucracy saw the need. Of course, that did not generally align with the local's needs. And, the task was done by hand.

The ad hoc snow clearing team. Front (l. to r.) Henri Schmetz, Sr., Jean Kevers, Henri Schmetz, Jr., Eugene Kevers. Rear (l. to r.): Jean Nyssen, Gerard Krafft, unknown.

An Offending Loaf Of Bread

As already noted, it was considered a capital crime to take food across the border that split Clermont, even if you had authorization to travel. One crisp, clear November afternoon in 1943, 10-year-old Marcel Schmetz and his friends were walking home from school along Rue du Bach. The boys approached the bridge over Berwinne Creek, from where, after crossing, Marcel would cut across the fields to his house. Suddenly, they stopped in their tracks. Several border guards were standing over the body of a young woman lying in the creek. She was motionless, face down in the water. The water just downstream was an odd pinkish color.

The guards' weapons were not shouldered, as was usual, and the young boys knew immediately what had happened. The young woman had attempted to sneak across the fields with what these guards considered contraband, apparently ignoring their orders to halt. Marcel and his friends would never forget the haunting image of that offending loaf of bread, bobbing in the creek.

The small bridge on Rue Du Bach over Berwinne Creek.
(The site where Marcel and his friends saw the murdered woman.)

Marcel's drawing of the border guards barracks,
550 meters (~600 yards) south of the bridge.

MORE ATROCITIES

By stark contrast, some of the "good" border guards were known to look the other way, especially for a bribe. Teenage brothers Albert and Guillaume Lemmens and their friend, Henri Hochtenbach, lived on the annexed side. They had friends and family on the occupied side who, due to the local rules, were chronically undernourished. The boys made a secret arrangement with a compassionate border guard for a nocturnal, humanitarian border crossing. On the agreed-upon night, the boys took a calf across the border at the planned time. Unfortunately, the crossing was observed by the guard's supervisor. The guard was executed on the spot. The three boys were arrested and taken away. They were imprisoned and tortured in Cologne. Later that year, on Hitler's birthday, they were beheaded. Marcel remembers that—between 1939 and 1945—on April 20, the Germans would decapitate or shoot prisoners as a birthday gift for Der Fuhrer.

A birthday gift for Hitler.

ATROCITIES UNDER GERMAN OCCUPATION
1940-1944
Aubel
RUE DE BATTICE
PUTTER FARM
HENRI-CHAPELLE MILITARY CEMETERY
MARCEL'S GERMAN SCHOOL
BELGIUM
La Clouse
3
WOMAN EXECUTED
Berwinne Creek
PICKLEPUSS CRASH SITE
2
Froidthier
RENÉ DEROUAUX KILLED
5
ELECTRIC TOWER
Henri-Chapelle
Crawhez
QUOIDBACH
SCHMETZ FARM
1
RENARD EXECUTED
REMEMBER 39-45 MUSEUM
LES BÉOLLES
La Minerie
GERMANY

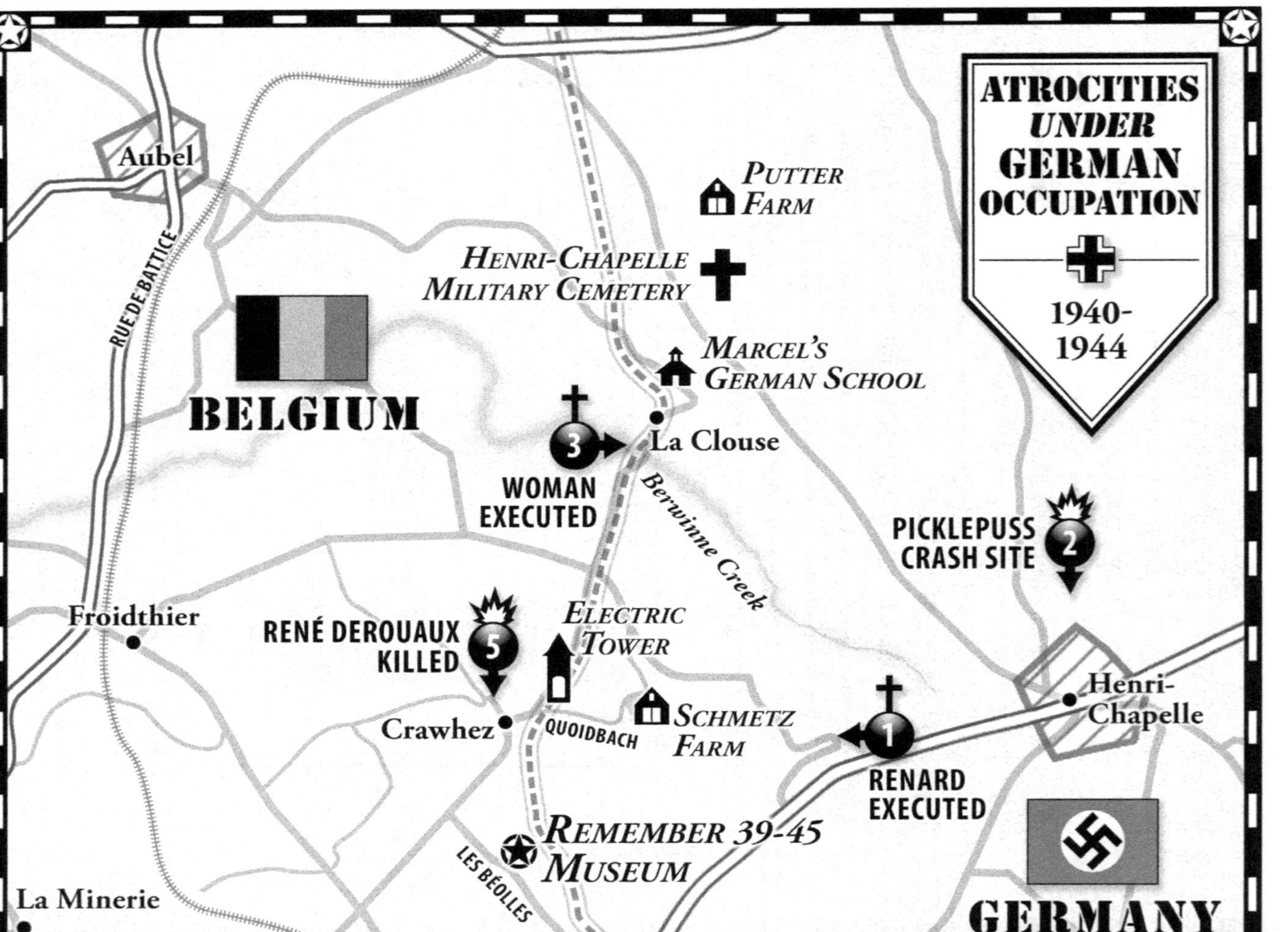

MAP 5

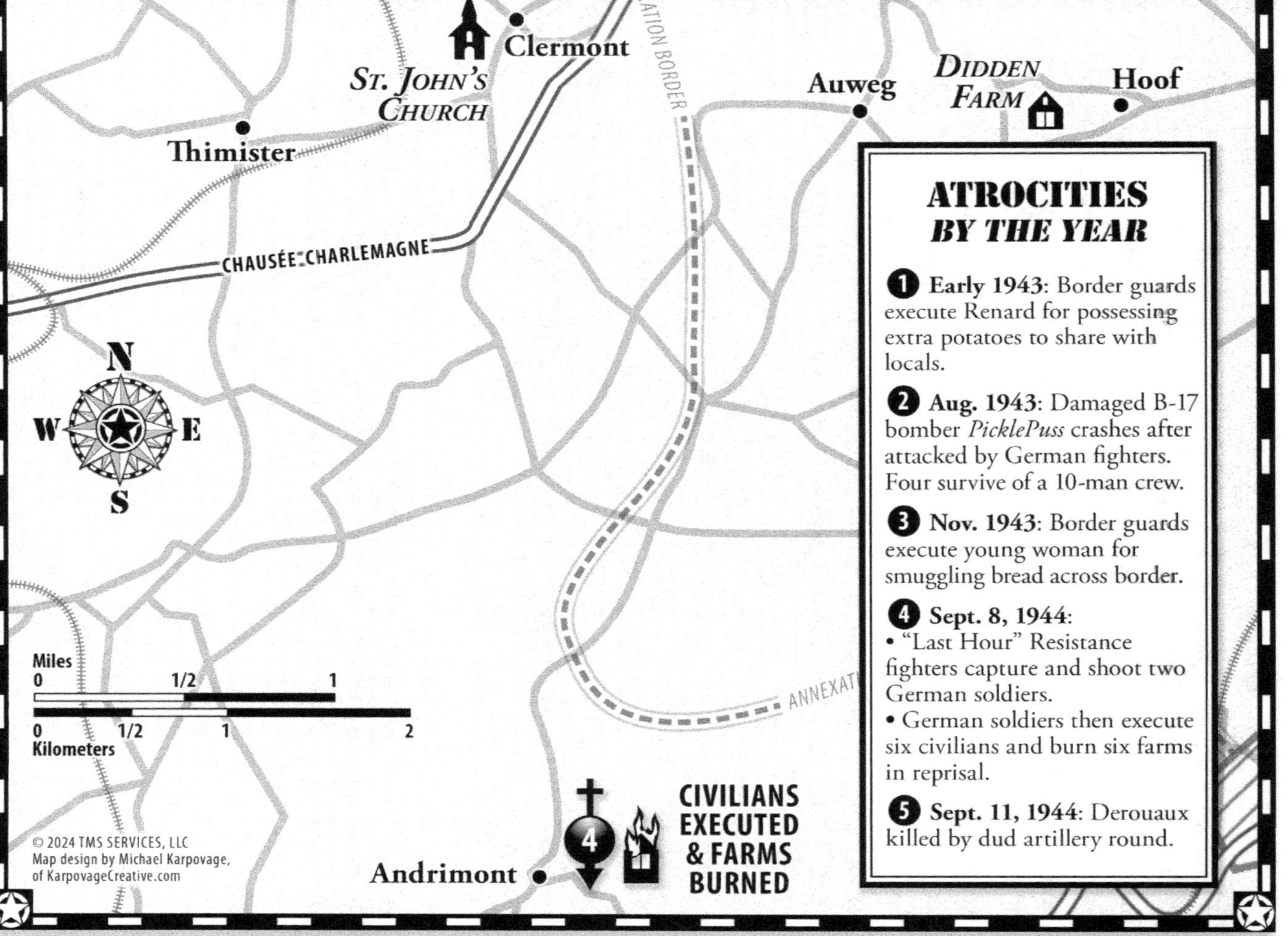
Clermont
St. John's Church
Thimister
ATION BORDER
Auweg
Didden Farm
Hoof
CHAUSÉE CHARLEMAGNE
N
W
E
S
ATROCITIES BY THE YEAR
1 Early 1943: Border guards execute Renard for possessing extra potatoes to share with locals.
2 Aug. 1943: Damaged B-17 bomber PicklePuss crashes after attacked by German fighters. Four survive of a 10-man crew.
3 Nov. 1943: Border guards execute young woman for smuggling bread across border.
4 Sept. 8, 1944:
• "Last Hour" Resistance fighters capture and shoot two German soldiers.
• German soldiers then execute six civilians and burn six farms in reprisal.
5 Sept. 11, 1944: Derouaux killed by dud artillery round.
Miles
0
1/2
1
0
1/2
1
2
Kilometers
ANNEXAT
4
CIVILIANS EXECUTED & FARMS BURNED
Andrimont
© 2024 TMS SERVICES, LLC
Map design by Michael Karpovage, of KarpovageCreative.com

Henri, Jr. Turns 17

In annexed Belgium, when the "new German" males turned 17, they were involuntarily conscripted into the German Army. Refusal meant prison or forced labor. In 1943, Henri Schmetz, Jr. became draft-eligible. Belgian men drafted into military service were generally sent to the Eastern Front to fight the Russians (even after the second front opened in 1944). The Germans thought self-preservation would be the ultimate motivation for them to fight well. Of course, desertion or surrender would lead to a firing squad at the hands of the ruthless Germans, or execution by the equally ruthless and vengeful Russian soldiers or civilians. For conscripted Belgians, there was a healthy chance you would not come home in one piece, if you even came home at all. Eight-thousand Belgians were drafted from the annexed area and sent to the eastern front. More than 2,200 never came home.

Dreading the day that Henri, Jr. would turn 17, the resourceful Henri, Sr. developed a plan. He constructed a secret space in their farm home. Though livable, it was cramped and windowless measuring approximately 4 feet by 10 feet. But it was cleverly constructed to minimize the risk of discovery by the Germans. Unbeknownst to Henri, Jr. before his 17th birthday, it was there he would hide from the Germans until Clermont was liberated.

In 1943, when that inevitable birthday came and Henri, Jr.'s German military orders arrived, his father sat him down to explain the plan. Young Henri was distraught, preferring instead to join the Resistance or go to England to join the fight. He did not want to hide, feeling ashamed he would not be doing his part for Belgium and worrying that people would think less of him. Fully knowing the ultimate fate of so many Belgian Resistance fighters—and their families—Henri, Sr. would have none of it. Torture and agonizing death were commonplace, sometimes even for uninvolved family members. From that day forward, Henri, Jr. would spend his days in that cramped space, venturing out only at night.

Henri, Jr. did not leave the house for 16 months. The Schmetz family knew they could not trust anyone to not report Henri, Jr. to the German authorities. To be discovered avoiding military service and disobeying orders would likely have meant a death sentence for Henri, his family and any co-conspirators, real or imagined. There were tangible rewards in those days for people who would betray a neighbor. So, nobody could know of the secret room. Even Gerard Krafft and his family never knew Henri, Jr.

was in the same building for all those months. Everyone believed Henri, Jr. was at the Russian front, fighting with the German Army.

"Where Is My Son?"

One day, a German Army representative came looking for Henri, Jr. His father was prepared. He told the representative that Henri had reported to the location on the date and the time prescribed in his orders. Then, Henri, Sr. strongly voiced his concern that the German Army did not know the whereabouts of his son. The German left, bewildered.

A week or two later a second army representative came knocking at the Schmetz residence. He again demanded to know where Henri was and why he had not reported for duty. This time, Henri Sr. feigned anger, looked the German in the eye and loudly exclaimed, *"Where is my son!? He promised me he would write to his mother every other day. I have not received one letter! Why do you not know where he is? He reported as ordered and you have lost him!"* Again, the agent left, muttering and shaking his head. And the Schmetzes went about their daily business.

Inevitably, there came a third visit. This time it was the Wehrmacht, inquiring as to Henri, Jr.'s whereabouts. Henri, Sr. became irate. *"I trusted you with my oldest son and you do not even know where he is! I have not heard from him since the day he reported to you. You cannot even tell me if he is alive or dead. I will look into this with the proper authorities!"* With that, Henri dialed up his charade, traveling to the town of Eupen, where the regional Nazi government and Kommandantür were located. Henri demanded they find his son immediately, even if he had been killed in action. *"How could the Army and the government not keep track of its soldiers!?"* Henri exclaimed. As a young soldier in Rhineland in 1919, and later as a prisoner of the Germans, Henri had learned an invaluable lesson. These Germans will respect you only if you look them in the eye and show no fear, weakness or hesitation. That lesson served him well, as the Germans never returned to the farm looking for Henri, Jr.

Loneliness And Isolation

Henri, Jr.'s existence was a lonely one. He had contact with ONLY his mother, father and little brother. And all of their lives depended upon the ability of 10-year-old Marcel to keep a BIG secret. While Henri, Jr. loathed

SCHMETZ
HOUSE
1ST
FLOOR
1944
Original hand-drawn floorplans
by Marcel Schmetz.
KITCHEN
MUD
ROOM
KRAFFT
FAMILY
HOUSE
DIVIDING WALL
STAIRS
DINING
ROOM
FRONT
DOOR
SUMMER
KITCHEN
DIVIDING WALL
KRAFFT
SCHMETZ
SUMMER
KITCHEN
FRONT
DOOR
PARKING
AREA
TOOL
SHED
© 2023 TMS SERVICES, LLC
Map design by Michael Karpovage,
of KarpovageCreative.com

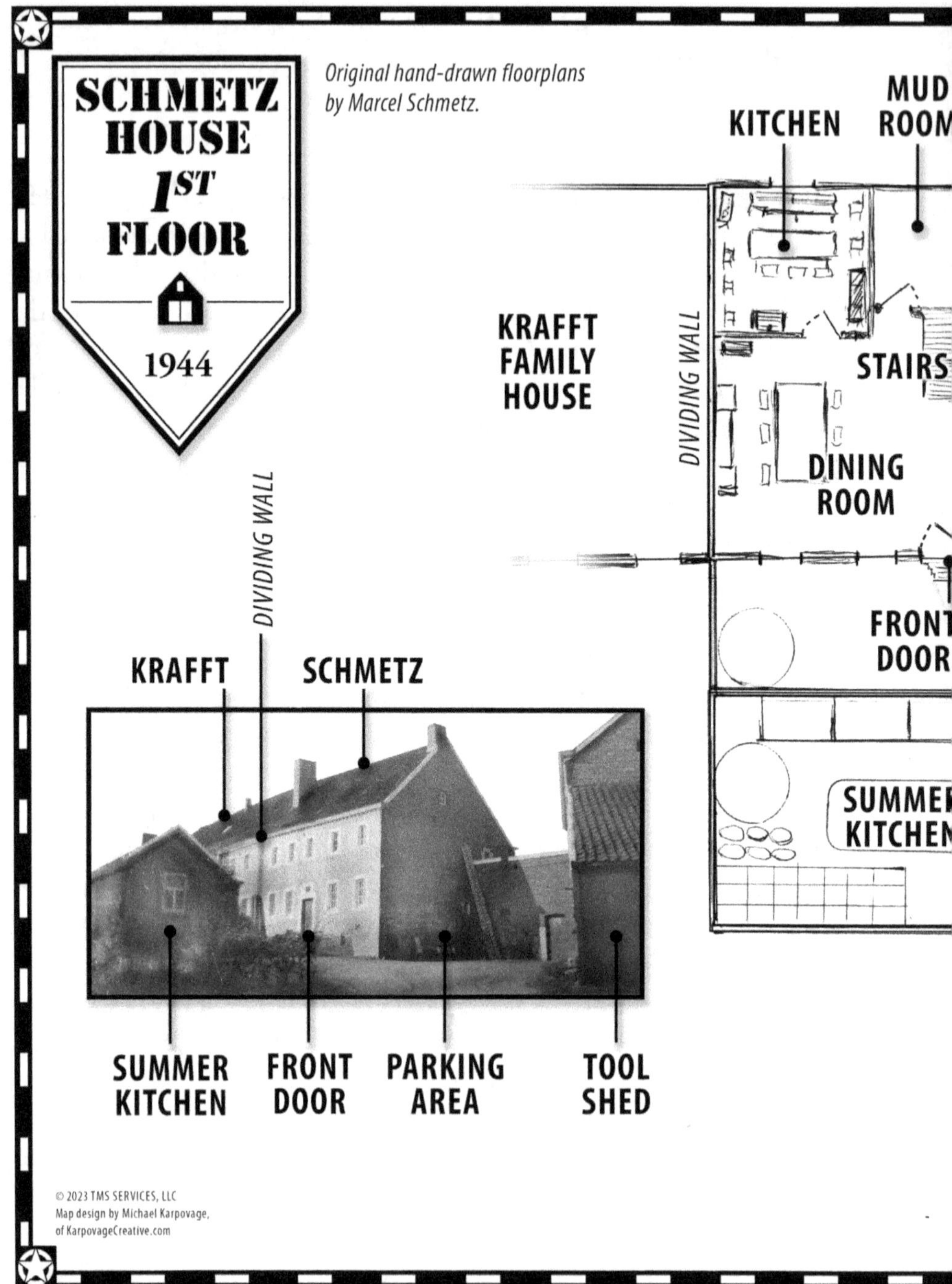

MAP 6

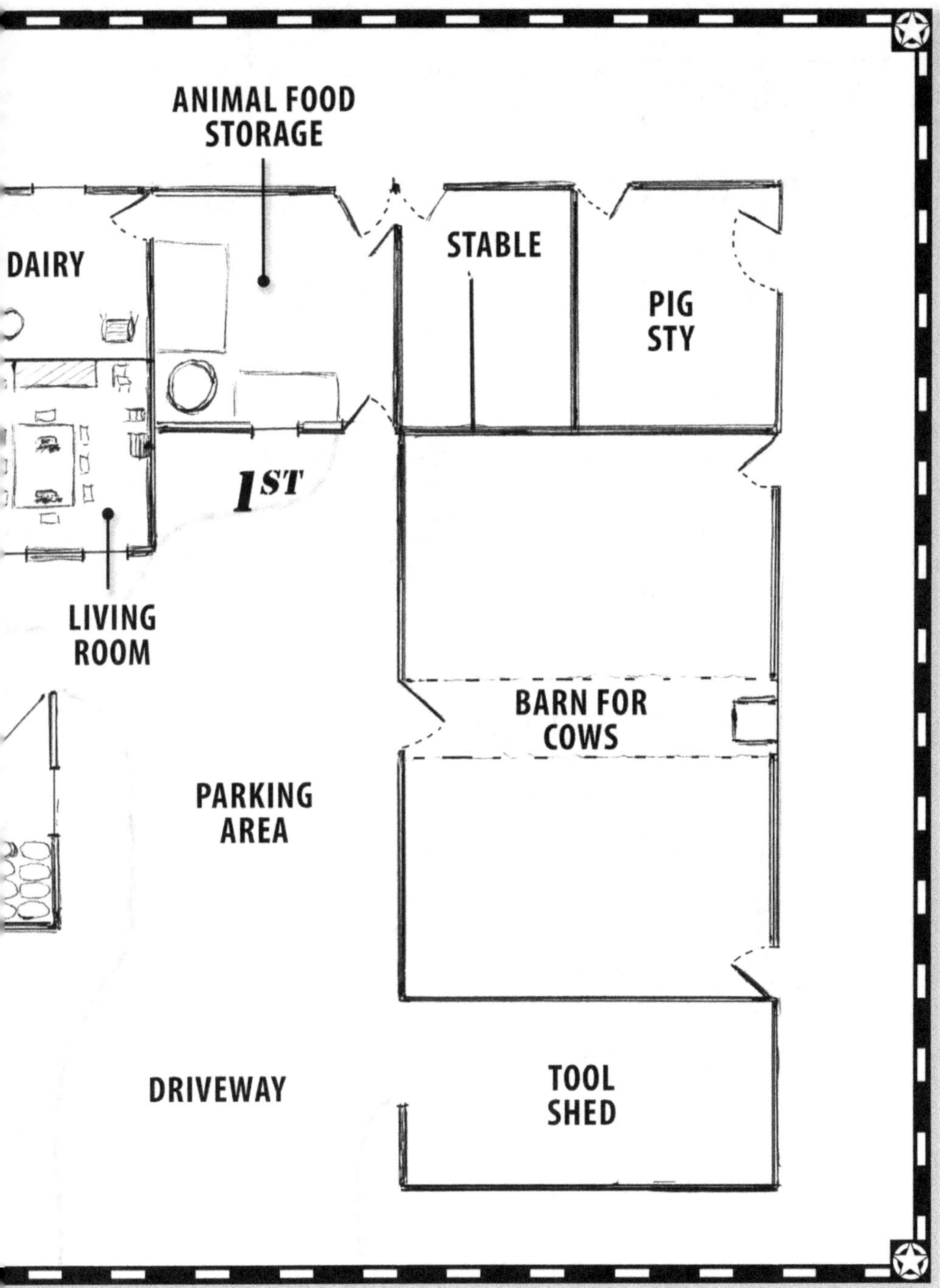
ANIMAL FOOD
STORAGE
DAIRY
STABLE
PIG
STY
1ST
LIVING
ROOM
BARN FOR
COWS
PARKING
AREA
DRIVEWAY
TOOL
SHED

Original hand-drawn floorplans by Marcel Schmetz.

KRAFFT FAMILY HOUSE

DIVIDING WALL

BED ROOM

STAIRS

HENRI JR'S HIDING PLACE

The bathtub border is essentially the footprint of the hiding place.

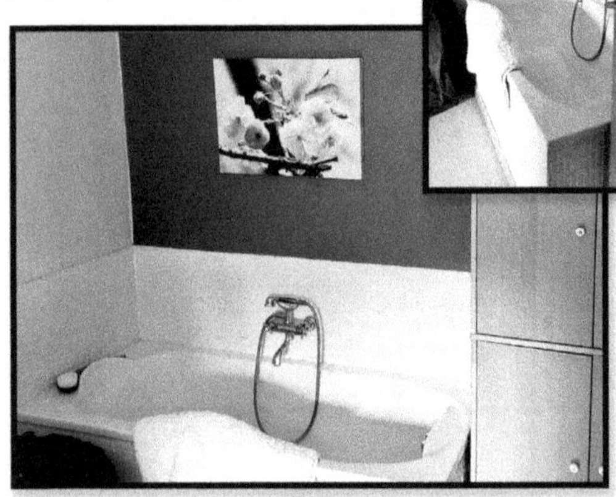

The current owner removed the faux walls of the secret room to enlarge it to a modern bathroom.

Map design by Michael Karpovage, of KarpovageCreative.com

MAP 7

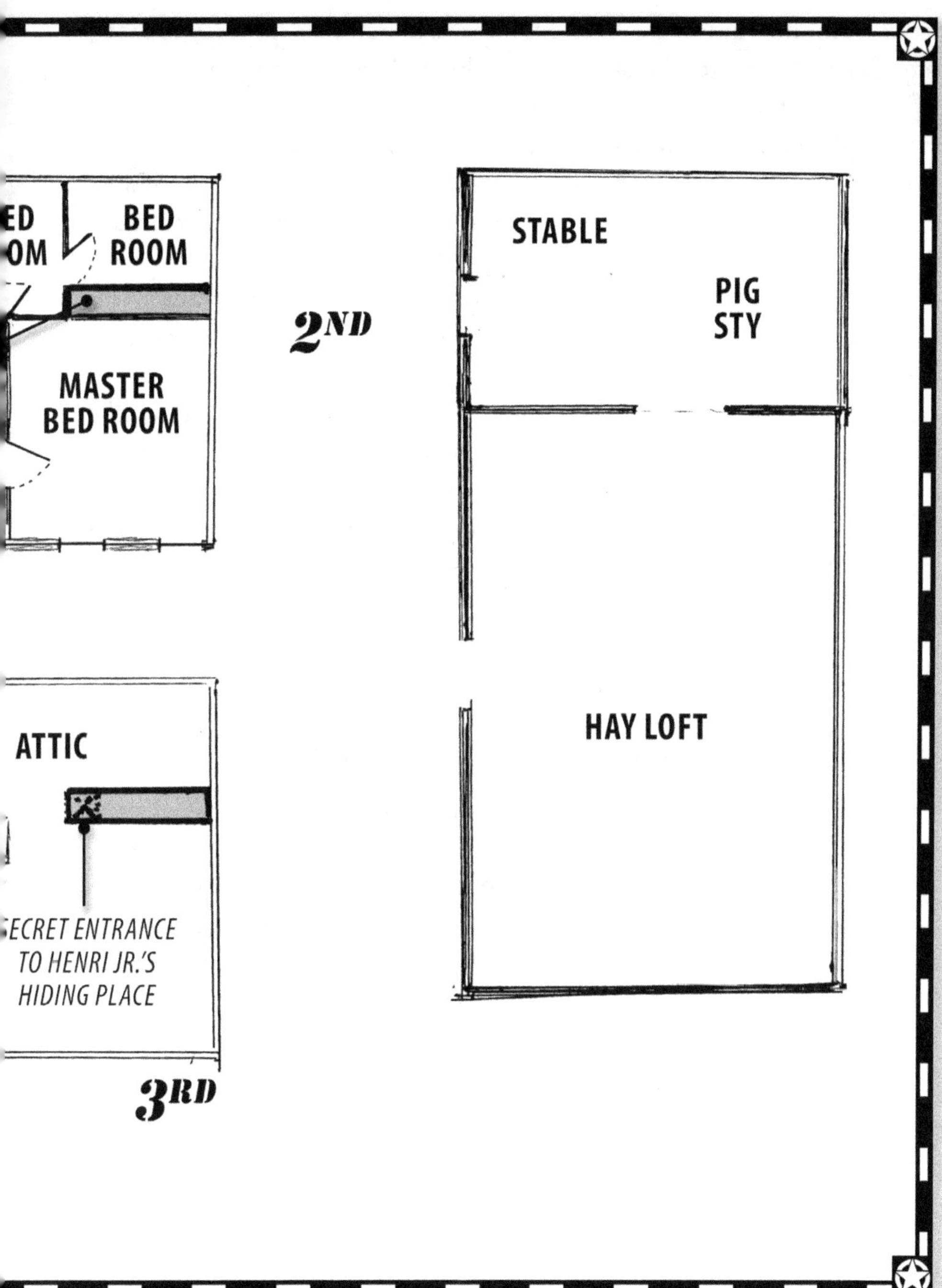
ED
OM
BED
ROOM
MASTER
BED ROOM
2ND
STABLE
PIG
STY
HAY LOFT
ATTIC
ECRET ENTRANCE
TO HENRI JR.'S
HIDING PLACE
3RD

his time in the windowless "box" his father had built, he did enjoy drawing and honed his sketching skills during his isolation. Even when he was "free" in the house, he had to be ready to bolt up to the attic and into his hiding place at a moment's notice, closing the trap door behind him. When possible, one of the family would scatter attic items over the trap door to make it harder to find if an unannounced inspection by the Nazi's or their surrogates occurred.

At night, to get fresh air, the teenager would sit in a darkened room near an open window. On rare occasions, without his parents' knowledge, he would sneak out to the barn for a few precious moments of freedom. Still, these seemingly innocuous jaunts carried huge risk for the family. Without notice and at any hour, German authorities would spot check a home, looking for contraband, illegal activity, escaped prisoners or deserters. Although the house had to remain unlocked, the barn could be locked to prevent theft. During the winter, Henri, Sr. would give the key to nearby border guards so, during their breaks, they could take refuge in the hayloft for a brief respite from the cold. Little did those guards know, but Henri, Jr.'s hiding place was just a few feet away.

A Haunting Sight

By the summer of 1943 the Americans and British were bombing Germany day and night, frequently with more than 1,000 bombers at a time. Though an amazing spectacle, it was so commonplace that it became background din for the Belgians who were under the aerial routes to the German targets.

One day in August, Marcel watched in horror as a lone, damaged and burning B-17 four-engine bomber limped by. Suddenly, it exploded, fell from the sky and crashed in the vicinity of the town of Henri-Chapelle, roughly a kilometer from where young Marcel stood. He ran to his house and told his father what he had seen. The two grabbed their bikes and peddled toward the smoke rising in the east. They rode into town and tried to navigate to the other side, but soon came to a roadblock that stopped them from getting any closer. From his vantage point, Marcel could see smoke and flames rising.

To see a "good-guy" bomber shot down was bad enough. But knowing the American crew may have perished was devastating. They were Americans trying to defeat the Germans and return his freedom. Marcel was heartbroken and would be haunted for years by what he had witnessed. ✪

3

BLOODY OMAHA

ON JUNE 6, 1944, BELGIUM HAD BEEN OCCUPIED OR annexed by the Germans for more than four years. The long-awaited invasion was finally underway, giving the Belgians—and the rest of Europe—hope that the German yoke would soon be removed. That day the Americans, British, Canadians and French landed at five separate beaches along the Normandy coast. The American 1st Infantry Division, the 29th Infantry Division, Ranger battalions, a separate tank battalion and many support units attacked the eight-kilometer (5 mi.) stretch of beach code-named Omaha. Though the landings were crafted with infinite detail, the historical adage, *"No plan survives contact with the enemy,"* a quote attributed to Helmuth von Moltke, 19th century German Field Marshall and war strategist, was on full display.

The 1st Infantry Division, known as the Big Red One (nicknamed for its prominent shoulder patch), had gained valuable combat experience in North Africa in 1942 and Sicily in 1943, before being pulled back to England for replacements, refitting and training for the D-Day landing. As did all infantry divisions at the time, the 1st Infantry Division consisted of three infantry rifle regiments: the 16th, 18th and 26th as well as numerous support units. Each regiment had three battalions: the 1st, 2nd and 3rd. Each battalion had four companies: 1st Battalion (A, B, C, D); 2nd Battalion (E, F, G, H), and 3rd Battalion (I, K, L and M). The Army did not use the letter J in its unit nomenclature as it could easily be confused with I. The first three companies in each battalion were rifle companies; the fourth was the heavy weapons company. A Regimental Combat Team was a Rifle Regiment augmented with other units, such as artillery, medical, engineer and/or tanks.

On D-Day, the Big Red One was in the first wave at Omaha Beach. As the initial landing crafts prepared to drop their ramps at 0630 hours, weather, beach obstacles and the sheer volume of enemy defensive fire scattered many of them significant distances from their designated landing points. In the chaos, Regimental Combat Teams, and even divisions, became intermixed. Some landing craft tried multiple times, unsuccessfully, to unload their cargo of men and weapons, and their landings were pushed back almost into the second wave. The 16th Regiment was in that first wave. The 18th Regiment was scheduled to land at 0930 hours, as the second wave, and the 26th Regiment would follow at noon in the third wave.

Bill Clements

Private Bill Clements, a 20-year-old from Saratoga Springs, NY, was a combat engineer with the 1st Engineer Combat Battalion of the 1st Infantry Division. He and his fellow engineers in Company A were assigned to support the 16th Regimental Combat Team in the first wave. Clements was tossed about the Landing Craft, Tank (LCT) as it navigated the choppy waters of the English Channel. The LCT was considerably larger than the Landing Craft, Vehicle and Personnel (LCVP, often called a Higgins boat), in which many of the first-wave troops were transported. Because of its size, an LCT could carry up to four M-4 Sherman tanks in addition to troops.

German artillery had begun firing at the assault wave while they were well offshore, and long before they were ready to unload. One shell screamed in, hit the small wheelhouse at the rear of Clements's craft and tore off the legs of the LCT's skipper. Clements heard the "peck, peck, peck" of German machine gun fire striking the wide steel ramp, which was still raised at the front of the landing craft. A second shell actually hit the ramp, and Clements recalls two men running to the back of the LCT with "their rear ends gone, their pants were blown off." Clements and three other GIs crouched behind a tank, hoping it would shield them. Just as it occurred to them that the tank was precisely what the Germans might be aiming at, a third shell struck the box for the LCT's machine gun ammunition, just behind them.

Two of the three soldiers next to Clements were killed instantly, the tops of their heads blown off. The third soldier was missing his left arm.

Clements was untouched. When the LCT hit the beach, the ramp was too damaged to lower. Clements and others began to jump off the back of the LCT, into more than four feet of water. In addition to his full combat load, Clements was carrying 50 pounds of TNT, intended for blowing up beach obstacles.

Suddenly, due to all the weight he was carrying, Clements was faced with the real possibility of drowning. He began to shed anything he had not already lost in the awkward jump overboard. After what seemed an eternity, Clements made it on to the beach. There, along with the other assault troops, he was pinned down by relentless mortar and machine gun fire.

Eventually he started to dig a foxhole. As he dug, Clements heard a voice booming through the din. "There are only two kinds of people on this beach, the dead and those who are going to die," the voice screamed. It was a colonel, encouraging the men to get off the beach. A small group of GIs had used bangalore torpedoes to blast a path through the barbed wire on the grassy area between the beach and the bottom of the cliff ahead of them. Clements and others began moving toward the gap in the wire and up the hill. Glancing back at the hole he had just left, Clements watched another soldier slide in, just as a mortar round scored a direct hit.

Clements had lost his rifle in the water, but found another on the beach. Later, a lieutenant told him to get rid of it. As it turned out, the weapon was a German rifle. Had the Germans found him with it, he would have been executed. Not to mention that his ammunition would not fit.

Bennie Zuskin

Private First Class Bennie Zuskin, from Hampton, VA, was 18 years old when he stepped off the ramp of Landing Craft, Infantry #401 (LCI) and into the water off Omaha Beach. He was part of the 1st Platoon of Company C, 26th Infantry Regiment of the 1st Infantry Division. His landing craft was in the third wave, which was planned to land around noon. For the most part, the first two waves were still crowded on the chaotic beach. There was not enough room for the third wave to land and the success of the invasion was far from a sure thing at this point. General Omar "Brad" Bradley, in charge of the invasion ground forces, delayed the third wave until around 1700 hours, not wanting to needlessly sacrifice men and materiel.

Thankfully, the first two waves rallied. The 26th Regiment was known as the Blue Spaders, because of the blue spade-like device on the regiment's distinctive insignia. Zuskin was assigned as an assistant gunner and ammunition carrier for his squad's Browning Automatic Rifle (B.A.R.) gunner.

Sam Messina

That gunner was Private First Class Sam Messina, from Lancaster, PA. Messina was the smallest man in the squad and he was carrying the largest weapon. The BAR weighed 16 pounds empty, more than 18 pounds loaded. By comparison, Zuskin's M-1 Garand rifle weighed nine pounds.

The LCI dropped its ramp on a sand bar. As the men advanced ahead, most disappeared into deeper water. Zuskin's first recollection upon leaving that craft was of seeing only Messina's hands and BAR above the water, as he tried to keep the weapon dry. Zuskin grabbed Messina's uniform and pulled his head above the water until they could make their way to more shallow footing. That was when they saw the hundreds of dead bodies, on the beach and floating in the shallow surf. Scores of wounded soldiers were clinging to beach obstacles as the tide slowly moved in to swallow them.

Zuskin saw the "shingle", a four-foot-high pile of rock where the sand met land. It was a much shorter run for the third wave as, by then, the low tide had advanced to high tide. As he scaled the shingle and headed for solid ground, a German shell struck nearby. The blast propelled a rock into the back of his loaded pack. Zuskin remembered thinking he had gotten his "million-dollar wound" and his war would soon be over. But there would be no such luck. Bruised, but largely uninjured, Bennie Zuskin would continue the fight for another 237 days without another scratch. Sam Messina would not be quite as lucky.

Rocky Moretto

Private First Class Rocco (Rocky) Moretto—C Company, 26th Infantry, 1st Infantry Division—was also aboard LCI #401. The 19-year-old kid from Hell's Kitchen, New York City came from a rough neighborhood. But it was nothing compared to the hell-on-earth he witnessed coming off that LCI's ramp on June 6, 1944. Like Sam Messina, Rocky went underwater shortly after leaving the boat. At the time, it was maintained by many of his buddies that the diminutive Rocky Moretto was actually shorter than his M1 Garand rifle. Now, the 60 pounds of equipment he was carrying was pulling him straight to the bottom. As the horrifying seconds ticked away, Moretto shed what he could and struggled onto the beach. There he saw many of his friends, dead or dying. The code name for his section of Omaha Beach was EASY RED. Muttering to himself, Moretto said, *"What joker came up with that?"*

Robert Blett and Jimmy Carr

Private First Class Robert Blett, from Worcester, MA, was inducted into the Army at Fort Devens, MA in February 1943. There he received his uniforms. Not long after, was put on a train to an unknown destination. He learned he was sent to Camp Pickett, VA where combat medics received their training.

Though not pleased about becoming a medic, Blett chose to make the best of it. After training, he and 30 other soldiers were shipped off to Fort Dix, NJ to await transfer overseas. "Hurry up and wait" became their mantra, as it was not until November that Blett learned he was going to England. While at Fort Dix, he had become good friends with PFC Jimmy Carr, from Lonsdale, RI. They had spent their weekends in New York City or Trenton, New Jersey, and took the same train home to visit their families, Jimmy getting off first in Providence, RI.

In December, Blett and Carr received their orders for England, crossing the Atlantic on the Queen Mary. After four days at sea, they landed in Scotland. From there they took a train to the replacement depot in England. All the men who had left Fort Dix were assigned to the 1st

Infantry Division at Tidsworth Barracks near Salisbury. They were a little more than 80 kilometers (49.7 mi.) from London. There the men were broken up and assigned to different regiments. But the luck of the two friends would continue. Both were assigned to C Company, 1st Battalion, 26th Infantry Regiment.

The two men knew the 1st Infantry Division was one of the most experienced divisions in the Army, having fought in North Africa and Sicily before returning to England. Going to war with combat veterans—rather than "rookies"—was very comforting for the two young replacement medics. Months of training followed, before it was finally time to board the ships that would take them across the English Channel. The night before, Blett went to Carr's compartment and they reminisced about their fun times together. They exchanged their parent's addresses and promised if one did not make it back the other would write his family.

Charlie Amero

Private First Class Charlie Amero was with D Company, 26th Infantry Regiment. It was the heavy weapons company for the 1st Battalion, and would be parceled out to A, B and C Companies, for a given operation, as the battalion commander saw fit. In addition to their personal weapons, Amero and his fellow soldiers were responsible for the heavy machine guns and 81mm mortars, which added significant firepower to the three rifle companies. Amero, from Essex, MA, hit the beach with one of his friends, PFC John Lebda, from Miller Run, PA. The first waves of the D-Day invasion were primarily combat engineers and infantry. The engineers' job was to clear the beach of obstacles, so the infantry and supporting vehicles could advance, and later waves would have an easier time reaching the beach. The infantry's job was to cover them as they did their job. Amero set up his machine gun near the bottom of the stone shingle where the sandy beach ended. He provided continuous covering fire for the rest of the company until they made it to dry ground. He and his assistant gunner then picked up their gun and moved it up the bluff to provide another layer of covering fire.

John Lebda

Private First Class John Lebda's first recollection of landing at Omaha Beach was devastating. As the landing craft ramp dropped, one of his "very best buddies" stepped off the end of the ramp and immediately sunk in the deep water. Private Charles Martin was considered the baby of the unit. Just 17 years old, the 120-pound teenager was weighed down with 60 pounds of equipment. He drowned before he could extricate himself.

By the time D Company landed, the tide had risen, and the remaining distance to the sand was relatively short. Lebda crossed the beach and made it to the top of the bluffs unscathed. His unit quickly advanced to the infamous hedgerows, where Germans were around every corner until proven otherwise.

In his memoirs, *A Million Miles To Go*, Lebda recounts an episode that churned his adrenaline. After a few days, the Company's vehicles caught up with the men. Each soldier had to take a two-hour turn on guard duty during the night to protect their fellow soldiers and the small motor pool that hauled their heavy weapons.

As Lebda recalled, after a few hours of pitch-black guard duty the mind sees and hears everything as a threat. Lebda heard "stomping and crunching" at the end of his stint and could only assume that the "Jerries were creeping up on me" There was a box of hand grenades in his foxhole. As the crunching noises came right up to the other side of the hedgerow, he began pulling pins and throwing grenades at the infiltrators. The explosions caused groans and thrashing noises. Some of his guys ran over and he told them "a bunch of Krauts were ready to attack so I blasted them."

Everyone was relieved when daylight finally arrived. Someone yelled, *"Hey Lebda, come and see all the Krauts you killed."* Spread before him was a herd of cows, "dead and belly up." For some time, he received the requisite "moo-moo catcalls and razzing."

The Graves Registration Company

An especially important unit, the 607th Quartermaster Graves Registration Company, was among the first groups in the third wave. It was their job to care for deceased military personnel. In theory, the deceased were to receive burials at large, temporary cemeteries far enough from the battlefield to allow their graves to remain undisturbed for the rest of the war. The third platoon of the 607th Quartermaster Graves Registration Company, consisting of 25 men, was assigned to go in with the 26th Infantry Regiment. They were transported in the same LCVP.

Vito Mastrangelo

Technical Sergeant Vito Mastrangelo, from Visalia, CA, was on that landing craft. The men made it to the beach and found cover in a shell hole. Before they could advance, a German sniper pinned them down. Periodically, shots rang out, kicking up dirt around their hole. The Graves Registration personnel were not trained as assault troops. Though they carried M-1 carbine rifles, their weapons were largely for self-defense, not tactical battlefield operations. So, they waited. Hours went by before several riflemen were dispatched to find and eliminate the enemy sniper. The sniper, who had kept the platoon pinned down for hours, turned out to be a French woman. No one ever really knew whether she was a Nazi sympathizer, or had just grown weary of the death and destruction the war had caused for her and her neighbors. She did not survive to answer the question.

The next day, Mastrangelo and his soldiers began collecting the dead on the beach. For Temporary Cemetery Number 1, the planners had chosen an area overlooking the beach. Unfortunately, that site was still under German control. Instead, they chose the far end of the adjacent landing site on the western-most part of Omaha Beach, code-named Dog Green. There they opened Temporary Cemetery Number 2 and began to bury the Allied dead.

At noon on June 8, they were joined by the Second Platoon. By the end of the day, they had buried 486 men near the beach. On June 10, Temporary Cemetery Number 1 was able to begin operations. A mere 10 days after the first wave of the invasion hit the beach, these two platoons had interred 1,510 Americans, 48 Allied soldiers and 606 of the enemy.

On June 7, the 4th Platoon landed on Utah Beach to support the 4th Infantry Division. There, another temporary cemetery was established. The 1st Platoon and Headquarters Platoon did not arrive until the end of June. The 607th Graves Registration Company remained on the beachhead for nearly three months, providing dignified care for those lost throughout northern France. Tech Sergeant Vito Mastrangelo and his men were exhausted. But their work had only begun. ✪

4

THE TIDE BEGINS TO TURN

AS THE SUMMER OF 1944 WANED, THE AMERICAN forces who had helped chase the Germans out of France were paused near the France-Belgium border. This was not due to any German defensive posture. It was because the Americans had outrun their logistics support. Fuel for thirsty armored vehicles could not be brought up fast enough or in large enough quantity to keep the forward line moving in unison. Combat engineers had built pipelines through France, but those required close guarding, as civilians would steal the precious gas, or soldiers would trade it for alcohol.

By September 1, 1944, the Big Red One was positioned in the Aisne region of France, northeast of Paris near the Belgian border. They would fight and advance on a general line from the Aisne region, through Belgium, toward Aachen. This put Clermont directly in its path. On September 2, the Americans crossed into Belgium and liberated Mons, a town just north of the French border. They steamrollered east to Charleroi, Namur, Liege, Verviers, Eupen and across the West Wall into Germany. And they made the 340-kilometer (211-mile) journey across Belgium in just 10 days. The 26th Infantry was on the right (southern) flank, the 16th in the middle and the 18th on the left (northern) flank with the 1st Engineer Combat Battalion supporting the advance.

As the division rolled through Liege, the soldiers were enthusiastically welcomed by ecstatic men, women and children welcoming them with flowers, sweets and schnapps. C Company of the 26th Regiment was greeted by a local policeman who spoke no English. He did, however, make it abundantly clear he wanted to join them to help fight the Germans. And he would not take "no" for an answer. Staff Sergeant Hugh Coltran, the

ADVANCE
OF THE 1ST ID
BIG RED ONE
1
D-DAY, JUNE 6
to NOV. 29,
1944
N
W
E
S
Thames R.
London
BRITAIN
Dover
Dunkir
Calais
Portland
Portsmouth
Boulogne
XX
1
ENGLISH CHANNEL
Cherbourg
Somme R.
Le Havre
OMAHA
BEACH
JUN 6
FRANCE
ST. JEAN
DE DAYE
JUL 20
Rouen
Caen
Saint-Lô
CAUMONT
JUL 13
NORMANDY
Seine R.
Oise R.
Falaise
Paris
LA FERTÉ-MACÉ
AUG 15
Chartres
© 2023 TMS SERVICES, LLC
Map design by Michael Karpovage,
of KarpovageCreative.com
SEINE RIVE
CROSSING AUG 2
MAYENNE AUG 8

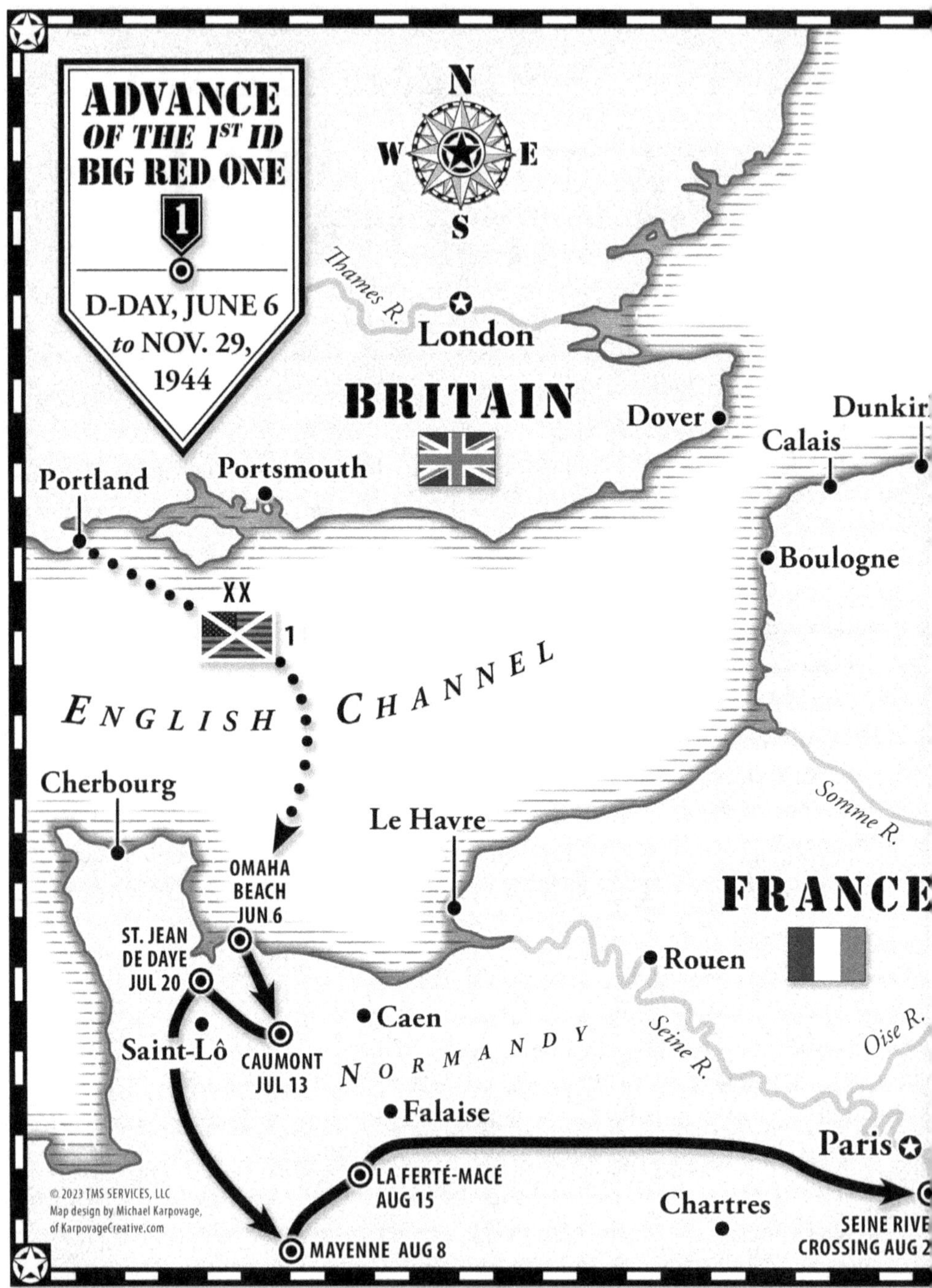

MAP 8

Amsterdam
NORTH
SEA
NETHERLANDS
WEST WALL
Meuse R.
GERMANY
Antwerp
AACHEN/STOLBERG/
HÜRTGEN FOREST
OCT-NOV
BELGIUM
Scheldt R.
MERODE
NOV 29
Brussels
Rhine R.
NÜTHEIM
SEP 13
MONS
SEP 4
LIÈGE
SEP 8
EUPEN
SEP 11
Namur
ARDENNES
MAUBEUGE
SEP 1
Meuse R.
WEST WALL
Mosel R.
Bastogne
LUXEMBOURG
SOISSONS
AUG 30
Aisne R.
Reims
Meuse R.
Marne R.
Miles
0 20 40 60
0 20 40 60 80
Kilometers

ADVANCE
OF THE 1-26TH
BLUE
SPADERS
SEPT. 1 to
SEPT. 13,
1944
NORTH
SEA
N
W
E
S
Brugge
Canal
Antwerp
Scheldt R.
Ghent
Leie R.
Escault R.
Brussels
BELGIUM
Louvai
Lille
MONS
SEP 4
Charleroi
Sambre R.
FRANCE
Bavay
MAUBEUGE
SEP 1
Escault R.
Amiens
Somme R.
Oise R.
Miles
0 10 20 30
0 10 20 30 40
Kilometers
III
26
1
© 2023 TMS SERVICES, LLC
Map design by Michael Karpovage,
of KarpovageCreative.com
Oise R.
Aisne R.

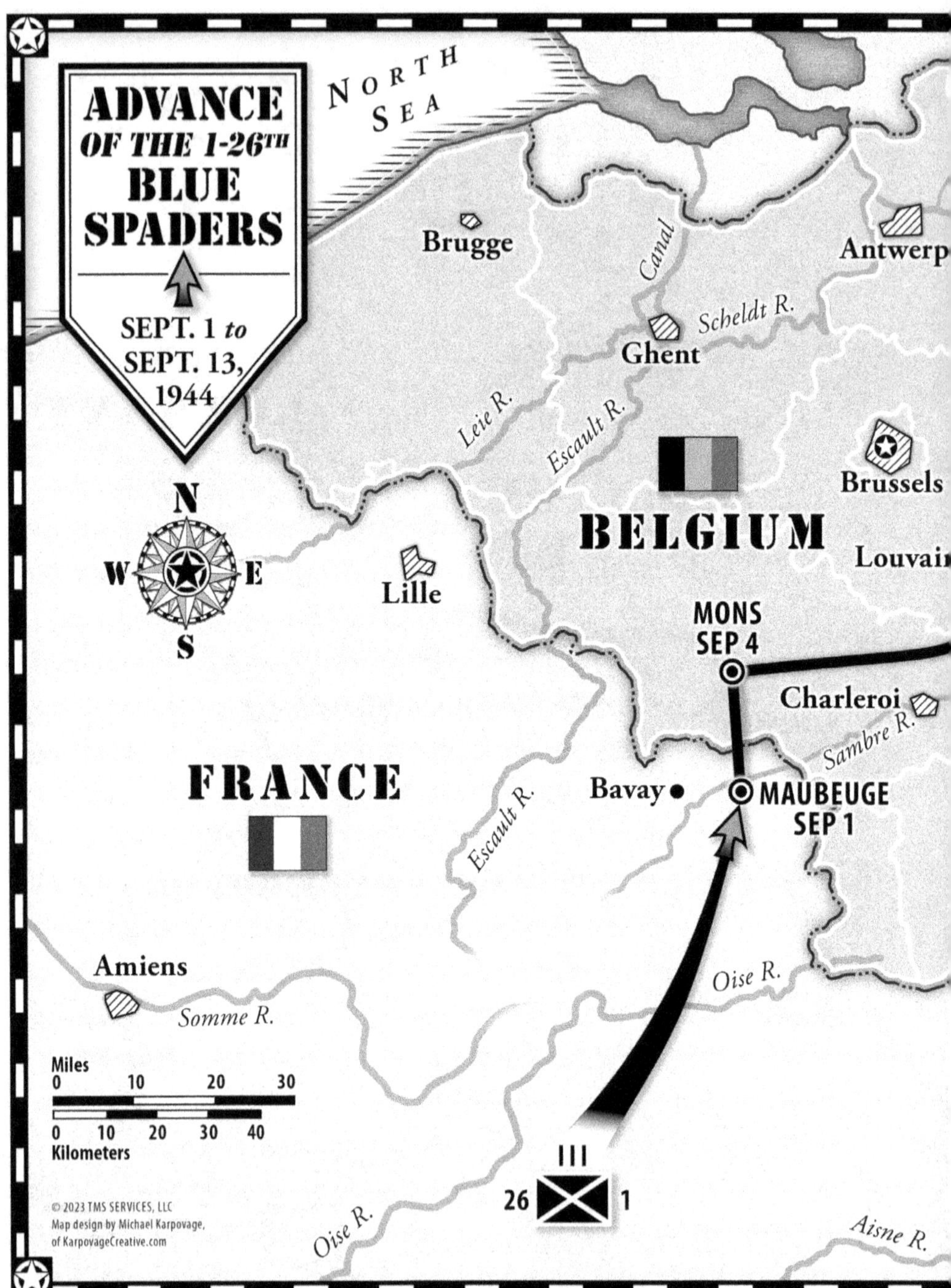

MAP 9

NETHERLANDS
Canal
Canal
Scheldt R.
Maas R.
WEST WALL
Rhine R.
München-Gladbach
Düsseldorf
Meuse R.
GERMANY
Cologne
Hasselt
Leuven
Maastricht
ANNEXATION BORDER
Roer R.
Erft R.
Aachen
Düren
Bonn
Rhine R.
CLERMONT
NÜTHEIM SEP 13
LIÈGE SEP 8
EUPEN SEP 11
VERVIERS SEP 10
Huy
Namur
Meuse R.
Ourthe R.
Mosel R.
Malmédy
WEST WALL
St Vith
Dinant
Meuse R.
ANNEXATION BORDER
A R D E N N E S
Our R.
Mosel R.
Bastogne
Bitburg
Bollendorf
WEST WALL
Trier
LUXEMBOURG
Arlon
Mosel R.
Meuse R.
Luxembourg

Second Platoon Sergeant, spoke passable French and helped to assimilate the Belgian into C Company.

"John The Belgique"

Coltran and his men embraced the newcomer. They found him a helmet, a rifle and even a Staff Sergeant uniform with the corresponding stripes on it. The combat veterans called him "John the Belgique." Initially, communication was awkward, but a "language" of equal parts English, French, German and pantomime was quickly concocted. John wasted no time convincing his new friends that he would be a valuable member of the team. Many of his family had been killed by the Germans, and his hate for them fiercely fueled his fire to fight them. He was willing to leave his wife and daughter behind to have the opportunity to exact some revenge. But John was not reckless. His knowledge of the countryside road nets and topography kept Moretto and the rest of C Company from learning lessons the hard way, which undoubtedly saved numerous lives. John was usually the first to volunteer for patrols, a dangerous job. Moreover, he wanted to be the point man on those patrols, an even more dangerous job. The information he gathered from the bodies of dead German soldiers—and from prisoners he captured—proved to be invaluable. John fought with C Company across Belgium and into Germany. After about two weeks of advancing against relatively light resistance, the company ran into an enemy buzzsaw. The Americans sustained heavy casualties and C Company had less than 50 men unscathed.

Rocky Moretto was lying in a ditch with a new soldier next to him, trying to avoid bullets and shrapnel. So many new replacements had been thrown into the fray that there was no way you could know their names. Rocky tried to break the tension, asking, "Pretty tough, huh?" The soldier had a puzzled look. It was then Rocky realized he did not speak English. A long-lasting bond was formed in that ditch.

C Company was integral to the capture of Aachen in October, before moving on to the Hürtgen Forest campaign in November. During the fighting, John was wounded and was evacuated back through the Army's medical system. "Staff Sergeant" John fooled the Medical Corps, healed quickly and found his way back to C Company.

Many weeks later, Stars and Stripes, an official military newspaper published to provide uncensored news from soldiers and for soldiers, told the tale of "John the Belgique." The article was brought to the attention of the 1st Infantry Division's Inspector General, who proceeded to rain down the full Divisional wrath on Captain Donald Lister, the Company Commander. Threatened with court martial, Lister was forced to send John home in March 1945. The "arrangement" had seriously violated the Geneva Convention. Had he been captured, John would have been treated as a spy and executed. So, after six months of arm-in-arm fighting with the Americans, John mustered up a gracious and tearful farewell for his combat-hardened brothers. His were not the only eyes filled with tears.

C Company in the snow near Bütgenbach during The Battle of the Bulge, January 1945. (l. to r.) Lieutenant Leon Kowalski, First Sergeant Gerald Waldo, Sergeant Bob Wright and John the Belgique.

The Germans Retreat

In early September 1944, the Schmetz family felt hopeful that good things were happening. Radio broadcasts informed them the Allies had landed in France in June, and were making substantial progress through France and into Belgium. The burning question was, *"When will they make it to Clermont?"* Henri and Maria spoke quietly of this, only in their home, as they did not wish to be perceived as traitors. The Germans were still fighting, but rumors were circulating that they were giving up ground in Belgium.

By the second week of September, the citizens of Clermont noticed an increasing number of German soldiers and vehicles moving east, back toward Germany. Unfortunately, the local food supply often went back with them, one last torment for the Belgians.

Maria And Marcel Go Shopping

On September 8, German units began taking over homes in Froidthier, an area of Thimister-Clermont, just two kilometers (1.2 mi.) west of the Schmetz farm. They were among many falling back from Liege to set up a new defensive line. An assault gun—an armored, self-propelled, turretless gun—was camouflaged near the edge of the village to ambush any advancing Americans. The next day, Marcel and his mother went into Henri-Chapelle, a much larger city just two kilometers (1.2 mi.) east of the farm. They were searching for groceries, as there were none in Clermont. Though they found precious little left on the Henri-Chappelle store shelves, Maria and Marcel were able to scavenge enough goods to feed the family for a few days.

As they were leaving Henri-Chapelle, they encountered a group of German soldiers resting by the side of the road. The soldiers had horses, as many vehicles lacking petrol were now being pulled by horses. Knowing full well that German soldiers were known to antagonize and pilfer from the hungry civilians, Maria and Marcel approached cautiously. These soldiers, however, let them go along their way. Marcel assumed they had already taken whatever they needed and were simply happy to be heading home to their own families. Marcel and Maria were relieved to make it home safely with their meager rations.

The Local Resistance Movement

At that time, the Belgian resistance movement was a menagerie of forces, not particularly well-integrated or coordinated. Some groups were well organized with a command structure and trained operatives. Others were isolated bands of opportunists. The latter were referred to as the "Last-Hour Resistance" by their fellow Belgians. These small groups, or individuals, did not resist while the Germans were in full control. With liberation imminent, they carefully observed the progress of the approaching Allies, acting only when it appeared there would be little risk. And for good reason. Brutal reprisals against civilians were not uncommon when the resistance killed German soldiers.

There were still German units in the western part of Belgium, near the West Wall. On September 7, 1944, two young German soldiers entered the tiny village of Andrimont, not far from Clermont. Though the fighting would continue for another eight months, it was clear to many German soldiers that the war was lost. It was also no secret that German prisoners of war were treated quite well by the Americans, especially if they were transported to the United States, where food was abundant. U.S. Army policy mandated that prisoners were to be fed at least as well as their captors. Faced with the threat of being transferred to the Eastern Front, desertion became quite a logical alternative. Many beleaguered German soldiers would sneak off, find an American unit and surrender.

It was never clear if the two young Germans who entered Andrimont that night were intending to surrender to the Americans, or if they were simply trying to work their way home to hide with family. But it was, however, fairly clear they were just looking for a safe place to hide and sleep. The story would later be recounted by Marie-Louise Faniel to her son, Georges Aalberts, who became friends with Marcel.

An Ill-fated Decision

Marie-Louise lived with her parents in a farmhouse just down the road from the village water tower. She recalled that, on September 8, the young German soldiers were spotted—and captured—by a handful of Last-Hour Resistance fighters. It was never clear what these townspeople were thinking, as they knew full well the Americans were close. Nevertheless, the resistance fighters marched the two prisoners down the farm road, behind the water tower and past the local cemetery. There the two young Germans were shot and left to die. A local woman discovered the wounded boys and called the mayor. The mayor called the local German unit, and an SS unit responded.

Georges walking down the farm lane where two German soldiers were murdered.

The SS, or Schutzstaffel, was the elite, black-uniformed, protective echelon of the Nazi Party. It was founded by Adolf Hitler in April 1925 as a small personal bodyguard, and grew in importance as the Nazi movement grew. Given immense police and military authority, the SS was known to be quite merciless.

The responding SS soldiers were led to the injured men. One was already dead, the other critically wounded. The squad leader asked the surviving soldier who had shot them. The soldier, suffering from serious wounds and likely confused by his location, pointed in the direction of a nearby farm. It was the wrong direction.

Satisfied, the SS officer shot the suspected-deserter dead, and he and his soldiers headed off for the implicated farm. On the way, they ran into two local boys, who had heard the gunshot, were curious and had decided to investigate. The soldiers butchered one boy on the spot, and took the other along with them. The SS then set fire to the first farm they came to. Looking out his window, a neighboring farmer saw what was happening, ran out the other side of his home and hid. At the next property, the SS took a man and his son prisoner, placing them with the first young captive.

Their farm, too, was set ablaze. At the next home, a father and son were trimming hedges. They too were forced to join the group of prisoners. All five were then marched back to the water tower. On the way, the SS torched five more farms and their respective houses, including that of Marie-Louise Faniel. Marie-Louise, her mother and aunt were forcibly removed from the home and left to watch it burn. At the water tower, the five innocent captives were executed.

A mere 96 hours before the Americans would liberate Andrimont and run the Germans out, a group of Last-Hour Resistance fighters had made the ill-fated decision to shoot two young Germans who were of no threat to their village. That choice left six townspeople executed and six farms destroyed. ✪

The Faniel House and barn before the war, and after the SS reprisal, September 1944.

5

LIBERATION

THE BIG RED ONE BEGAN THEIR LIBERATION OF Belgium through the city of Mons, taking it on September 4. There they caught a large German force by surprise and inflicted heavy losses. More than 120,000 prisoners were taken. Liege was the next prize on September 8. Soon it would be Clermont's turn.

On the morning of September 11, Lucien Ernst, a farmer in Froidthier, left his home with a milk jug, disguising his real intentions. Ernst headed to the American headquarters in Herve to notify them of the German positions, especially that hidden assault gun set up for an ambush. At 1330 hours, armed with Ernst's information, the Americans began a one-hour artillery barrage on the newly identified targets. An infantry assault on Froidthier followed, and fierce fighting ensued.

By mid-afternoon Laurent Veeschkens, Clermont's town secretary, had heard rumors that Americans were in the area. He hung the Belgian flag inside the main window of the Clermont Town Hall. Soon, a single American jeep cautiously entered Clermont, along Rue René Rutten. Veeschkens asked the first soldier he saw if he was American. The soldier answered with the colloquial American, *"Yeah,"* not the more formal, *"Yes,"* that the Belgians understood. Panic immediately consumed every citizen within earshot, and someone quickly yanked down the Belgian flag. It took the bewildered American a few moments to realize the Belgian townspeople thought he had uttered, *"Ja,"* the German word for *"Yes."* The misunderstanding was quickly corrected, a collective sigh of relief was exhaled and the flag was reinstated.

LIBERATION
OF THE LIÈGE
PROVINCE
BY THE 1ST ID
1
SEPT. 8 to
SEPT. 16,
1944
NETHERLANDS
Fort Eben-Emael
Obsinnich
Visé
Fort Aubin-
Neufchateau
Aubel
Albert Canal
Meuse R.
Fort
Pontisse
Froidthier
SCHMETZ FARM
XX
1
Fort
Barchon
Clermont
III
18
1
Battice
Thimister
Fort
Evegnée
Fort
Battice
Herve
III
16
1
LIÈGE
SEP 8
Andrimont
Fort Fléron
Dison
III
26
1
BELGIUM
(LIÈGE PROVINCE)
Fort Chaudfontaine
Fort
Embourg
VERVIER
SEP 10
Pepinster
ANDOUMONT
SEP 9
Fort Tancrémont
© 2023 TMS SERVICES, LLC
Map design by Michael Karpovage,
of KarpovageCreative.com
Louveigne
Theux

MAP 10

WEST WALL
Aachen
Eilendorf
WEST WALL
Stolberg
Brand
Gemmenich
GERMANY
BÜSBACH
SEP 16
Preuswald
omburg
NÜTHEIM
SEP 13
Hauset
HÜRTGEN
FOREST
Eynatten
Henri-
Chapelle
Busch
B 26
WEST WALL
Welkenraedt
D 26
A 26
Rötgen
KETTENIS
SEP 12
C 26
imburg
EUPEN
SEP 11
ARDENNES
N
W
E
S
ANNEXATION BORDER
Monschau
Miles
0 1 2 3 4 5
0 1 2 3 4 5 6 7
Kilometers

PFC Bill Clements, Staff Sergeant John Tait, their platoon leader, Lieutenant Valerie W. Kosorek and three soldiers from the 1st Engineer Combat Battalion (PFC Sam McFalls, PFC Clark and Tech Sergeant Edwin Carmichael) were the liberators in the jeep that had advanced along Rue René Rutten. The townspeople swarmed the jeep along that narrow street and dubbed the combat engineers, "The Liberators of Clermont." In no time, the soldiers were being celebrated with cookies and schnapps, a moment photographed by townsman Joseph Baguette.

Below photo: Lt. Kosorek (Platoon Leader) on left with map; Tech Sgt. Carmichael on lower right in driver's seat; PFC Sam McFalls upper right. PFC Clements, with rifle between legs, is cut off at the top.

Above photo: The advance party of the 1st Engineer Combat Battalion rolls in to liberate Clermont, September 11, 1944. (By contrast to America, 9/11 is a very happy occasion in Liege Province.)

That photographic moment was poignant, rare and extremely fortunate, given that cameras were strictly forbidden by the Germans. Having an unexpired roll of film was even more amazing. At great risk to himself and his family, Baguette had kept his camera hidden for more than four years. Two months later, in a stunning coincidence, Baguette would recognize Clements in Spa, Belgium, while Clements was on respite from the front lines of the Hürtgen Forest. Baguette gave Clements a copy of the photo he had taken that morning on the Rue René Rutten. In a fascinating twist, Baguette's photograph would surface once again decades later.

Belgium's September 11

For the jubilant citizens in that rural part of Belgium, September 11 was a joyous day, a date still celebrated reverently to this day. (Almost six decades later, that date would become infamous for the American liberators and their descendants.)

As joyful as the people of Clermont were, the day for which they had all been waiting was not entirely happy. A day earlier, the Americans had been shelling the retreating Germans. After surviving more than four years of occupation, René Derouaux, a 16-year-old friend of Marcel's, was struck by an American artillery round while standing outside his home. The shell did not detonate, but René was killed instantly. Had it exploded, more of his family would have been killed.

As the Allies raced across France toward Belgium in 1944, German army units occasionally practiced manning their pre-built trenches. The locals also noticed SS units drilling in the nearby woods. At 2:00 am on September 11, 1944, back in nearby Hoof, Alice Didden heard vehicle and troop movements. She wondered if the liberators might finally be there. Her hopes were quickly dashed when she heard German voices. Suddenly, the trenches and woods were filled with German soldiers.

Alice's father had heard that bombs would not explode if they landed in piled hay, so he quickly evacuated his family to the barn. Four German soldier were already there. All of them could hear nearby fighting, closing in. At 11:00 am, Mr. Didden thought it would be safer in the basement of the house. The family made a run for it, with the German soldiers following closely behind. Around 1:00 pm, Alice's older sister smelled smoke, and their father stole a peek outside. A German tank had been hit and veered right into the barn. Their former hiding place was now ablaze. The family implored the soldiers to get the tank out of the barn and let the horses loose. But they refused. Suddenly, American warplanes strafed the tank, the barn and the surrounding woods.

It was not long before the Diddens again heard tanks clanking outside. But this time, they were American tanks, driving down the main road and plowing across the fields and trenches. Mr. Didden went out to greet them, informing them there were four German soldiers in his basement. The US infantry, not wanting to be caught in an ambush, led him to the basement with a rifle pressed to his back. Quietly, the Germans surrendered and were trucked off to a prisoner-of-war collection point. At long last, the Didden family was free again.

The fickleness of war is something to behold. While the citizens of Clermont were enjoying cookies and schnapps, their neighbors—less than 2 kilometers (1.2 mi.) away—were dodging bullets and shrapnel.

The 26th Regiment of the Big Red One formed the southern flank of the Division as it crossed Belgium and enjoyed the "easier" end of that fickleness gamut. They arrived in a calm Verviers on September 11, a day after liberating Liege. C Company was given a day of rest and bivouacked in a city park. The locals jubilantly swarmed them showering them with thanks.

A gentleman named Henri Lemaire struck up a conversation with Rocky Moretto, inviting him and three of his buddies to the Lemaire home to celebrate properly. Private First Class Warren Coffman, Sergeant Bob Wright and a third soldier joined Moretto on a short walk to 107 Rue de Dison. It was a large house and the occupants appeared to be rather wealthy. Bing Crosby records and refreshments were brought out and the soldiers danced with several young ladies, including the two Lemaire sisters, Lucienne and Josette.

After that treatment, it was hard to go back to the encampment. PFC Coffman picks up the story:

> ***"Several days later C Company was given another day of rest. We were not too far from Verviers. Without permission, Rocky Moretto and I hitch-hiked back to the city. We returned to see the family that befriended us. They invited us to stay all night so we accepted. We were given a nice big bedroom with feather beds. The room was completely blacked out and it was pitch dark. We had no trouble sleeping on those big, soft feather beds after sleeping on the ground all summer. We had planned to get up early the next morning and return to our unit, but no one woke us.***
>
> ***"When we got up, it was already noontime. We wanted to leave right away but they insisted we have breakfast first. Their son had gotten up early that morning and rode his bicycle to a farmer's house where he obtained fresh eggs, milk and***

some potatoes. It was too good to pass up so we stayed and had a delicious breakfast. I had heard an old First Division saying that it was all right to go AWOL (Absent Without Leave) as long as you were going towards the front.

"We were going towards the front but were having difficulty getting a ride. Finally a chaplain came along in a jeep and picked us up. He was suspicious of us at first, and thought we were going AWOL. We insisted we wanted to go back to our unit. He knew where our unit was located and took us there. We arrived just as the unit was moving out. Our buddies thought we went AWOL and had turned our equipment in to the supply truck. We quickly retrieved our gear and weapons and ran to catch our unit. It was a stupid thing for us to do but it turned out OK, and we sure enjoyed the bed and breakfast."

Henri, Jr.: Never The Same

When the liberation of Clermont began, Henri Schmetz, Jr. found himself feeling anxious about the disruption to his "normal" solitude. He had spent 16 months in maddening concealment, with no outside contact. Now, with strangers appearing all about, he was positively terrified to leave his sanctuary. Nearly two days passed after the Americans had liberated his town before Henri, Jr. could be coaxed from his hiding place. He emerged, trusting no one. Although Henri, Jr. never really discussed his true feelings with his family, Marcel knew he had difficulty making the behavioral adjustment after being isolated for so long. Marcel also believes his brother experienced survivor's guilt, as so many of his fellow Belgian teenagers had died fighting the Russians or resisting the Nazis. To Marcel, his big brother was never the same. Today it is called Post Traumatic Stress Disorder (PTSD).

MAP 11

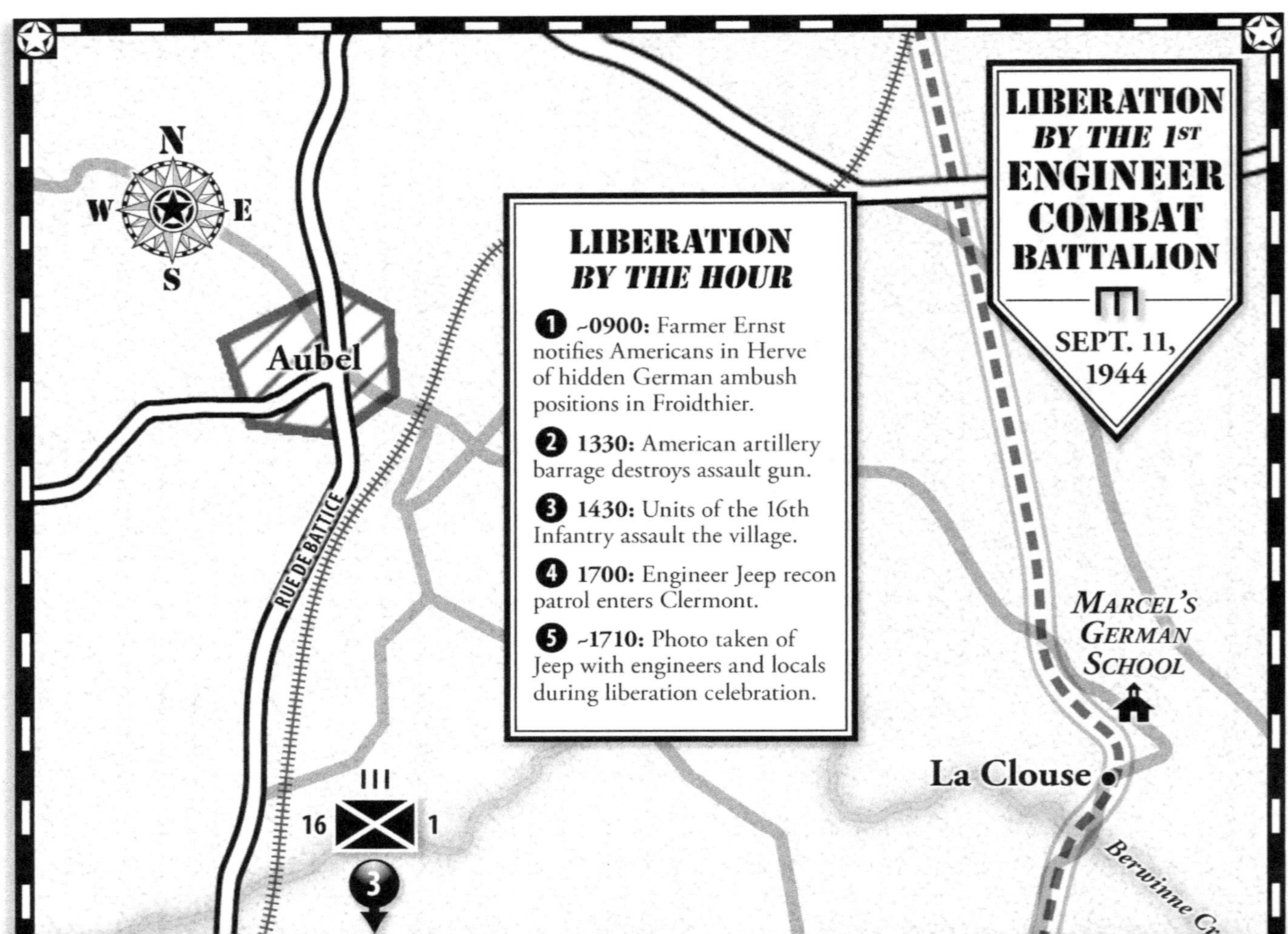
N
W
E
S
Aubel
RUE DE BATTICE
LIBERATION BY THE HOUR
1 ~0900: Farmer Ernst notifies Americans in Herve of hidden German ambush positions in Froidthier.
2 1330: American artillery barrage destroys assault gun.
3 1430: Units of the 16th Infantry assault the village.
4 1700: Engineer Jeep recon patrol enters Clermont.
5 ~1710: Photo taken of Jeep with engineers and locals during liberation celebration.
LIBERATION BY THE 1ST ENGINEER COMBAT BATTALION
SEPT. 11, 1944
MARCEL'S GERMAN SCHOOL
La Clouse
Berwinne Cr
III
16
1
3

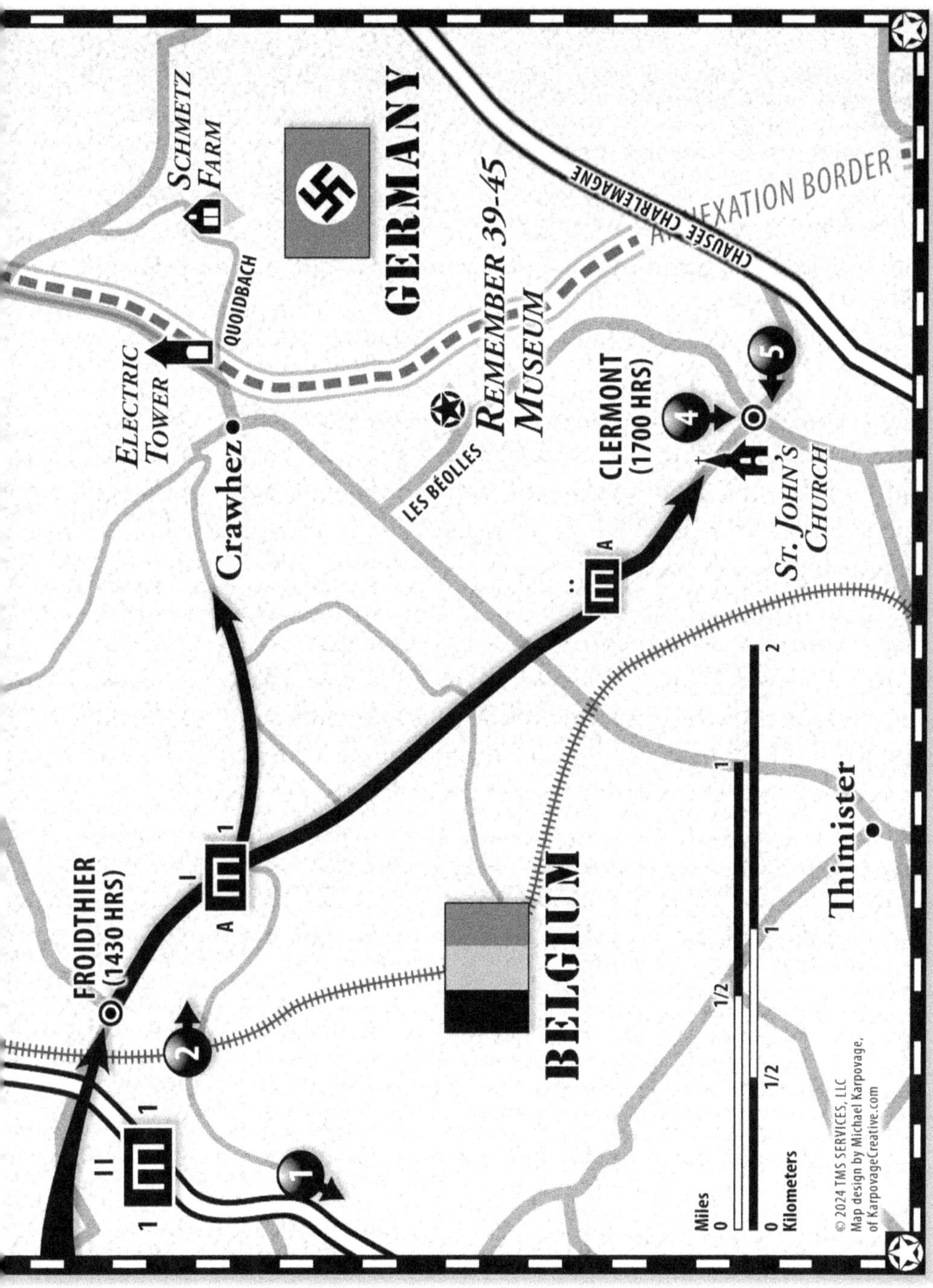
GERMANY
BELGIUM
Schmetz Farm
Quoidbach
Electric Tower
Crawhez
Remember 39-45 Museum
Les Béolles
Clermont (1700 HRS)
St. John's Church
Froidthier (1430 HRS)
Thimister
Chaussée Charlemagne
Annexation Border
Miles
Kilometers
© 2024 TMS SERVICES, LLC
Map design by Michael Karpovage,
of KarpovageCreative.com

Enjoying Freedom

For the Belgians, healing began slowly, especially in terms of nutrition. Compared to the German occupiers, the Americans were well supplied and well fed. This meant the endless confiscation of meat, vegetables and dairy products by the Germans had finally come to a merciful end. As life quickly improved for the Belgians, the Americans continued their push toward Aachen. Shortly, it would become the first major German city to fall to the Allies.

Other units of the Big Red One, particularly the 1st Battalion, 26th Infantry Regiment, rolled through the neighboring towns of Thimister, Eupen and others. Though momentous for the Schmetz family and their small town, it was just another of the numerous small villages the Americans had liberated since landing in Normandy three months before. Another tiny speed bump on the road to the vaunted Siegfried Line, now just 16 kilometers (9.9 mi.) away.

Everyone enjoyed their newfound freedom, especially the children. Schools were closed because the roads were dangerous. Thousands of American military vehicles were moving forward at high speed, even on the country roads. Army Quartermaster units chose Clermont as a fuel depot, which meant fuel trucks added even more traffic to the area. Hedges, fences and telephone poles were destroyed as tanks and half-tracks drove across the country. This meant considerable repair work for farmers and municipalities, but they felt it was a very small price to pay for their liberty. ✪

6

THE BRUTAL FIGHT CONTINUES

THE BIG RED ONE, WITH PFC BILL CLEMENTS AND his fellow engineers, was racing toward Germany. From Eupen, they moved east to the German town of Roetgen, swinging south of the first big prize of the European Theater: Aachen. It was from Aachen where Charlemagne—Charles the Great—had ruled the region around 768 A.D. Under the rule of Charlemagne, hailed as the "Father of Europe," most of Western Europe was united for the first time since the classical era of the Roman Empire. This included parts of Europe that had never been under Frankish or Roman rule.

The American 30th Infantry Division swung north of Aachen in a classic pincer movement, a military maneuver in which forces simultaneously attack both flanks of an enemy formation in order to envelop them. C Company of the 1st Battalion, 26th Infantry Regiment was part of the southern pincer. Robert Blett, in the 1st platoon, and Jimmy Carr, in the 2nd platoon, the two C Company medics who had met at Fort Dix, traversed the Atlantic together and hit the beach at Omaha, had spent the past three months caring for countless wounded soldiers. Once the Germans had been chased out of the Normandy hedgerows, their defenses began melting back across France, then Belgium. Now, with their homeland under attack, their resolve and resistance stiffened significantly. They sprang a counterattack, with artillery and tanks, on C Company and its platoons.

Blett's platoon leader, a Lieutenant Brooks (first name unknown), called for a bazooka team to move up the road and try to kill a tank approaching their position. The two-man team got close but were cut down in a roadside ditch by the tank's machine gun. Blett waited for the tank's turret to swivel around, looking for another target, and moving its cannon and machine gun away from the downed bazooka team. As the turret turned, Blett made a beeline for his comrades in the ditch. After running about 18 meters (~20 yds.), he watched the big gun swing back toward him. As the cannon fired, Blett dove into the ditch, unscathed, and crawled to the two men. One was dead. The other, PFC Reilly (first name unknown), was wounded in both legs. To bandage Reilly's wounds, Blett had to rise up ever-so-slightly. Each time he did, the tank's machine gun opened up. When Blett was almost finished controlling the bleeding, he noticed the tank pulling back. After withdrawing about 70 meters (75 yds.), the tank stopped. Suddenly, it charged forward, firing all its guns down the road at the platoon.

Reilly, the bazooka man, screamed, *"Grab the bazooka and get that SOB before he kills us all."* The bazooka was already loaded with a rocket. Blett picked up the weapon, raised it to his shoulder, took aim and pulled the trigger. As he recalled, *"It made a huge roar. The rocket hit the tank on the track and the side. The track was blown off and the tank started to smoke. The crew was trying to get out, but they were cut down by rifle fire from behind us."* Looking back, he saw Lieutenant Brooks, Sergeant Clayton Goode and the rest of the 1st Platoon charging up the road. Blett continued tending to Reilly as the 2nd Platoon moved up.

It was then someone jumped in the ditch to help. It was Blett's good friend, Jimmy Carr. The two medics splinted Reilly's legs, hailed a team of litter-bearers and loaded him on a jeep for a ride to the aid station. Blett hoofed it up the road and rejoined his platoon.

Robert Blett received the Bronze Star with Valor for his actions. It was not exactly how he had imagined spending his 21st birthday. As it turned out, the medic belonged to a very small, elite group of one: "Medic Tank Killer."

The Hürtgen Forest

Throughout October, the Big Red One fought in and around Aachen. November brought large numbers of new, inexperienced soldiers to the division to replace the killed, wounded and missing. On November 16, the Big Red One was turned southeast and thrown into an ill-conceived operation to clear the Germans from the Hürtgen Forest. They joined several other divisions, including the 28th Infantry Division, known as the "Bloody Bucket" for its red keystone patch. The meat grinder that was the Hürtgen Forest had been killing and maiming soldiers from both sides since

September 19. The foes fought back and forth in a 140-square-kilometer (54-sq. mi.) forested area, only five kilometers (~3 mi.) east of the Belgian-German border.

November 23 was another miserable day in the forest, the dismal skies pelting the 26th Infantry with a steady, cold drizzle. C Company was hunkered down in their foxholes, murmuring their usual G.I. affectations to Mother Nature. Bob Blett was eating cold C Rations with Sergeant Goode. As the rain pinged off their helmets they heard someone yelling that there were turkey dinners at the command post.

"There goes another poor bastard over the edge," Blett thought. There had been more than a few soldiers who had succumbed to "battle fatigue" in the three weeks they had been living in muddy holes with daily artillery barrages. But then, Lieutenant Brooks crawled over to their hole and proclaimed, *"It's Thanksgiving!"* As Blett recalls, each team took turns crawling from their protective foxholes to the command post. *"We had the works, Blett remembers, "turkey, mashed potatoes, squash, dressing, cranberry sauce, biscuits, coffee and mince pie for dessert."* The following day, the 1st and 2nd Platoons of C Company were consolidated to form a full strength platoon.

Blett writes:

> ***"We pushed off around noon through mortar, machine gun and small arms fire towards the town of Merode. We made it about a mile when we ran into heavy machine gun and rifle fire at a roadblock. Word came back that some men were hit up ahead. Jimmy and I made our way along the ditch near the road till (sic) we reached them. There were two men hit, one in the leg and the other in the arm. We patched them up as best we could, and they told us that Lieutenant Brooks was also hit and was down about 50 feet in front of us at the side of the road. We got to the Lieutenant and saw that he was hit in the chest. We decided to carry him back to where the other two wounded men were. When we got back we could see he was turning grey. I got out a plasma unit and we inserted it in his arm. I asked the***

> ***man who was wounded in the leg if he could make it back on his own. He said he could make it and started back. We asked the man wounded in the arm if he could hold the plasma unit while Jim and I carried the Lieutenant. We hadn't gone far when I heard this screaming "88"shell coming in. Instinct told us to drop the Lieutenant and hit the ground. I never heard the shell go off, just one blinding flash and the sensation of floating thru (sic) the air and then darkness. The next thing I knew I was in the Battalion Aid Station. My right eye was completely shut and my head and face were badly swollen. My arm was throbbing and burning. I was moved to the rear of the Aid Station where the Catholic chaplain gave me the Last Rites of the Church. That was the last thing I remembered."***

Bob Blett woke up in a Field Hospital 15 miles behind the front lines. He was to be transported to a General Hospital in Liege for definitive surgery. As he was awaiting further evacuation, a fellow platoon member, Joe Risco, was carried in with a wound that required surgery. He also carried the news that Jimmy Carr, Lieutenant Brooks and the soldier holding the plasma had all been killed in the blast.

While Blett was recuperating in Liege, he found the address his friend Jimmy had given him on D-Day. Here he picks up his story:

> ***"I was never very good at writing letters and had a hard time trying to think of the right things to say (to Carr's mother). I tried to tell her what happened. I didn't want to tell her he had his arm, shoulder and half his head blown off. It took me a couple of hours to write a page and a half, it was a very hard thing to do."***

Six months in combat—especially the previous two—had seriously depleted the Big Red One's combat power. Those months had been particularly bloody and costly, with heavy casualties for little gain. From November 29-30 alone, the Big Red One had more than 300 men killed in action, in addition to numerous wounded, captured, missing and non-battle casualties. Now, as November turned to December, the 1st Infantry Division was withdrawn from the Hürtgen Forest, and dispersed to the west to the northern shoulder of the Ardennes Forest. It was time to reconstitute and refit. Again. With young, inexperienced men fresh from the States.

Simultaneously, the 28th Infantry Division was relocated to the southern end of the Ardennes, spread thinly to cover a front more than 16 kilometers (~10 mi.) long, and directly opposite the Germans. As a general rule, a division would defend an 8-13-kilometer front (5-8 mi.). But it was winter, the coldest one in decades. U.S. leadership considered the hilly, heavily forested Ardennes region to be a "low-risk" location for the battered divisions. After all, they reasoned, who would attack in such severe conditions, on terrain that was unforgiving for armored vehicles, even in clear weather?

The Liberators Return

On Monday morning, November 27, a jeep drove up the road to the Schmetz farm in Quoidbach. Marcel remembers the day clearly. Mondays were laundry day for his mother, and the visit interrupted her chore. Two American officers hopped out of the jeep and approached Henri, Sr as he stepped out of the barn to see what was going on. Politely, the officers asked if four of their soldiers could stay at the Schmetz home for R&R (rest and relaxation). *"Of course!"* Henri replied. He told the officers his family would do anything to help the Americans. Maria quickly cleared the first-floor living room, and set it up to sleep four. Then, the Schmetzes waited patiently for their special guests to arrive. And waited. And waited some more...

Early evening came and went, and no one showed. Henri and Maria were becoming frustrated. Many of the day's chores had been neglected, as they wanted to look their best to warmly greet "their" soldiers. Around 8 pm, the soldiers finally arrived. But there were not four of them. There were 10!

All the soldiers began to set up their cots in the living room that had been meticulously prepared to hold four. *(See Schmetz Floor Plans – Maps 6 and 7).* Within minutes, Maria and Henri heard more trucks, jeeps and other vehicles approaching their farm. The entire D Company, 1st

Battalion, 26th Infantry, 1st Infantry Division had come to bivouac on the Schmetz farm!

By then, it was dark, cold and raining. An officer approached and asked Henri if there was any way he could find more space for the soldiers. Maria insisted they stay in the house, and hurriedly moved Henri Jr.'s and Marcel's beds into their parents' bedroom. They cleared as much furniture and belongings as they could and, before long, dozens of cots filled every available space. But there were still more soldiers outside in the rain. Henri, Sr. approached an officer, asking if it would be OK for some of the soldiers to sleep on hay in the barn, even though it meant sharing space with cows.

Henri did not realize that for battle-weary infantry soldiers—used to sleeping in frigid, muddy foxholes,—the barn was the epitome of five-star luxury. But even the large barn could not accommodate the remainder of the company. So, they removed all the farm equipment from the smaller outbuildings, and spread straw on the floor. That space, too, was quickly filled, meaning some of the men had to sleep in their vehicles. For one day and one night, the Schmetz farm had become home to more than 200 American soldiers.

D Company soldiers snuggling with the cows.

When daybreak dawned, the still-displaced Americans were able to find other farms that could house them. When it was all sorted out, 110 D Company soldiers were domiciled on Marcel's parents' farm. Dispersed across the property, the enlisted men made themselves comfortable in various rustic lodgings. Officers and non-commissioned officers stayed in the house. It would take many farms and towns in and around Clermont to accommodate the entire Division of roughly 15,000 soldiers. The 26th Infantry Regiment alone was scattered from Aubel in the north to Verviers in the south.

For weeks, the soldiers had been living in their uniforms, the filthy, grimy clothing becoming a second skin. In a few days, the Quartermaster unit caught up with D Company and the soldiers were given new uniforms. The discarded uniforms were piled in a back field and, mercifully, burned. Marcel always regretted not saving a few of them.

"Official Business"

Although enjoying their well-earned respite from war, the soldiers still made time to wash their own uniforms and clothing. Maria was ever grateful for the kindness the soldiers heaped on her family. So she would iron each of the uniforms after they had been washed. Some of the soldiers noticed that Maria used an old-fashioned iron that needed to be heated frequently on the hot stove. One morning, the soldiers drove to Aachen on "official business." They had sacrificed much in blood and souls for the conquered city. Most of the civilians had evacuated before that attack and much of the city was devastated, so it was relatively unpopulated. Many of the homes were still furnished and full of belongings. The way the soldiers looked at it, anything they found that was useful was nowhere close to payback. Later that day, the small party of soldiers returned to the farm. With them were a half-dozen "liberated" electric irons for Maria.

The Schmetz family had only one radio in their home. Three months earlier, the Nazi propaganda rantings of Josef Goebbels and Adolph Hitler were the only broadcasts they could receive. Now, the airwaves were filled with wonderful music and real news. Unfortunately, based on the accommodations arrangement, only officers and non-commissioned officers could listen to the one radio. But, after a few more "official business" trips to Aachen, the countryside around the farm was permeated with swing, jazz and big band tunes. The soldiers—and many of the locals—loved it. Henri, Sr. was not so sure. He called it, "wild music."

One day, much to the chagrin of the soldiers, one of the newly-acquired radios quit working. Henri, Sr. rode his bicycle to Aubel to see Mr. Belly, the only radio repairman in the area. Both returned on their bicycles and Mr. Belly was taken to his "patient." It took only minutes before the radio was blaring "wild music" again. The happy soldiers asked Mr. Belly how much they owed him. *"Nothing,"* he replied. *"You are our liberators!"* One of the soldiers asked him to sit and stay with them for a little while. Meanwhile, they quietly passed a helmet around for donations. Soon it returned full of paper bills. Marcel recalls that Mr. Belly "almost had a heart attack," as he had never seen that much money before.

In addition to again having their own milk, butter and foodstuffs, items previously confiscated by the Germans, the Schmetz family could now enjoy the cooking and baking from the American field kitchen, which the soldiers generously shared. For the entire time Maria was a host, she never cooked a single meal. The field kitchen was set up across from the front door of the house. It was in the existing "summer kitchen," the small outbuilding where food was cooked during the sweltering summer months, in order to keep the living area cooler. There was no space for indoor cooking, anyway, as bitter cold temperatures kept everyone huddled inside the house, barns and outbuildings. In addition to the men, the house and farm were also packed with weapons and equipment.

The Schmetz family congregated in the kitchen for meals and indoor activities. The D Company Commander, Captain Walter Stevens, and his command staff would join the family for meals. There was a bench passed down from Henri, Sr.'s father that was placed along the back of the dinner table. The officers sat there, and it was christened, "The Bench of Honor."

D Company had wonderful cooks who gratefully served the family two hot meals every day. Breakfast was a veritable feast of milk, buttered pancakes, sausage, powdered eggs and bacon, some smoked and some fresh. After a while, a disconcerting problem arose on the farm: the Schmetz's pigs all became rather ill. Marcel wryly recalls that any leftovers were given to the pigs, who promptly ate like...pigs. As it turned out, after surviving four years on very little food, the pigs' gluttony wreaked havoc on their delicate porcine digestive systems.

Marcel recalls another marvel. The Company cooks did not have to bake bread. There were special units known as U.S. Army Quartermaster Bakery Companies. They used special trucks and were 100% mobile. Each company baked 60,000 pounds of fresh bread each day. A truck came by the farm daily and delivered large paper "bread sacks." On each bag was printed, "Return When Empty." The following day each came back full again. Marcel was quite amazed by this fascinating magic trick.

Mobile bread bakers.

After all they had endured, the Schmetz family loved and appreciated everything they were served. They considered the Army field kitchen cooks to be master chefs. Marcel, his brother and their mother especially came to love a concoction the Americans called "peanut butter." Never had the three of them tasted anything quite like that sticky, tasty treat. But, as Marcel recalls, Henri, Sr. was a bit "picky" when it came to the new delicacy. Henri, it seems, did not much fancy how the peanut butter stuck to the roof of his mouth. And, according to Marcel, old farmers do not eat what they do not know.

Masterpieces And Souvenirs

PFC Charles (Charlie) Amero was a budding artist who made numerous drawings while on the farm. Henri, Jr., who enjoyed playing the accordion, but could not make music during his "imprisonment," had taken up drawing to pass the time and keep his mind occupied during those long months. Charlie and Henri, Jr. quickly bonded, encouraging each other, critiquing their work and complimenting each other's "masterpieces."

Drawing by Henri, Jr.

John Lebda recalls the little blonde boy who was so enamored with the soldiers who blanketed his farm. He and his fellow D Company men were exhausted from months of fighting, but Marcel had enough energy for them all. If Marcel could not coax a soldier into playing a game with him, he was just as happy to watch him clean weapons or perform other soldier chores.

Marcel clearly remembers PFC Clayton Ramaly doing a lot of the cow-milking and pitching in with Henri, Sr. on other farm chores. He and some of the other men would periodically return to Aachen, again on "Official Business." After the city fell the previous October, there were many "souvenirs" to be had, but no one really had the inclination to acquire them, let alone lug them around a battlefield. Now, however, with the promise of up to four weeks of rest, some of those souvenirs would come in quite handy. Ramaly picked up a pot that he could use to cook potatoes and onions. But his real specialty was making popcorn, much to Marcel's delight.

One day during the second week of December, Major General Clift Andrus, the Commanding General of the First Infantry Division, came to visit the troops. Glancing from soldier to soldier, Marcel's family knew immediately something special was happening. The men were bathed and shaved, all weapons and equipment were cleaned and no food was laying around the farm. (The leftovers had been given to the pigs, whose stomachs by then had rallied quite nicely.) The soldiers lined up and stood at attention before General Andrus. After addressing his soldiers, the General came to speak with the family. Communication was awkward, as the General spoke no French and the family spoke no English. Marcel realized General Andrus was thanking his family. The General went from one farm to the next, profoundly thanking each for hosting his soldiers. No one imagined then that only a few days later, a major German counterattack would shatter their peace, striking fear in the hearts of the Belgians that the Germans would return again. ✪

7

THE BATTLE OF THE BULGE

AT 0530, DECEMBER 16, 1944, GERMANY'S BORDER with Belgium and Luxembourg exploded. More than 2,600 artillery pieces began pummeling known and suspected American command-and-control positions. Hitler's last-ditch, grand counteroffensive in the west had begun. The Oberkommando der Wehrmacht (OKW)—or Wehrmacht High Command—code-named it Unternehmen: Wacht am Rhein (Operation: Watch on the Rhein). Its intent was to infer a defensive posture, while keeping their real offensive intentions secret. It was the operation the contemporary American press has coined, "The Battle of the Bulge." It would last almost six weeks and involve more than one million soldiers.

As artillery was relentlessly shelling the border and beyond, three well-equipped German Armies rolled into the snowy Ardennes region. It was the same heavily forested region U.S. leadership had previously considered a "low-risk" location for the battered 28th Infantry Division. The Germans plowed into only four American divisions, two of which had no combat experience: the 106th Infantry Division, which had replaced the veteran 2nd Infantry Division just 48 hours earlier and the 99th Infantry Division whose nom-de-guerre was "Battle Babies." At H-hour, the time the attack began, more than 200,000 Germans faced off against roughly 83,000 Americans along a 96-kilometer (59.7 mi.) front. At some points along the line, the Germans had as much as a 10-to-1 manpower advantage.

The unexpected and ambitious German objective was to race west to the Meuse River, pivot north and capture Antwerp. Hitler thought this would effectively split the British and American armies. He hoped this would crack the western alliance and cause one, or both, of his enemies to sue for peace.

BATTLE
OF THE
BULGE
ATTACK
DEC. 16 to
DEC. 25,
1944
Maastricht
NETHERLA
BELGIUM
Liège
XXXX
1
HODGES
Namur
Meuse R.
Huy
Sambre R.
Charleroi
Ourthe R.
Hotten
Meuse R.
Dinant
GERMAN LINE DEC 25
N
W
E
S
A
R
D
Bastog
Meuse R.
Neufchâteau
Miles
0
10
20
0
10
20
30
Kilometers
© 2023 TMS SERVICES, LLC
Map design by Michael Karpovage,
of KarpovageCreative.com
FRANCE
Semeis R.

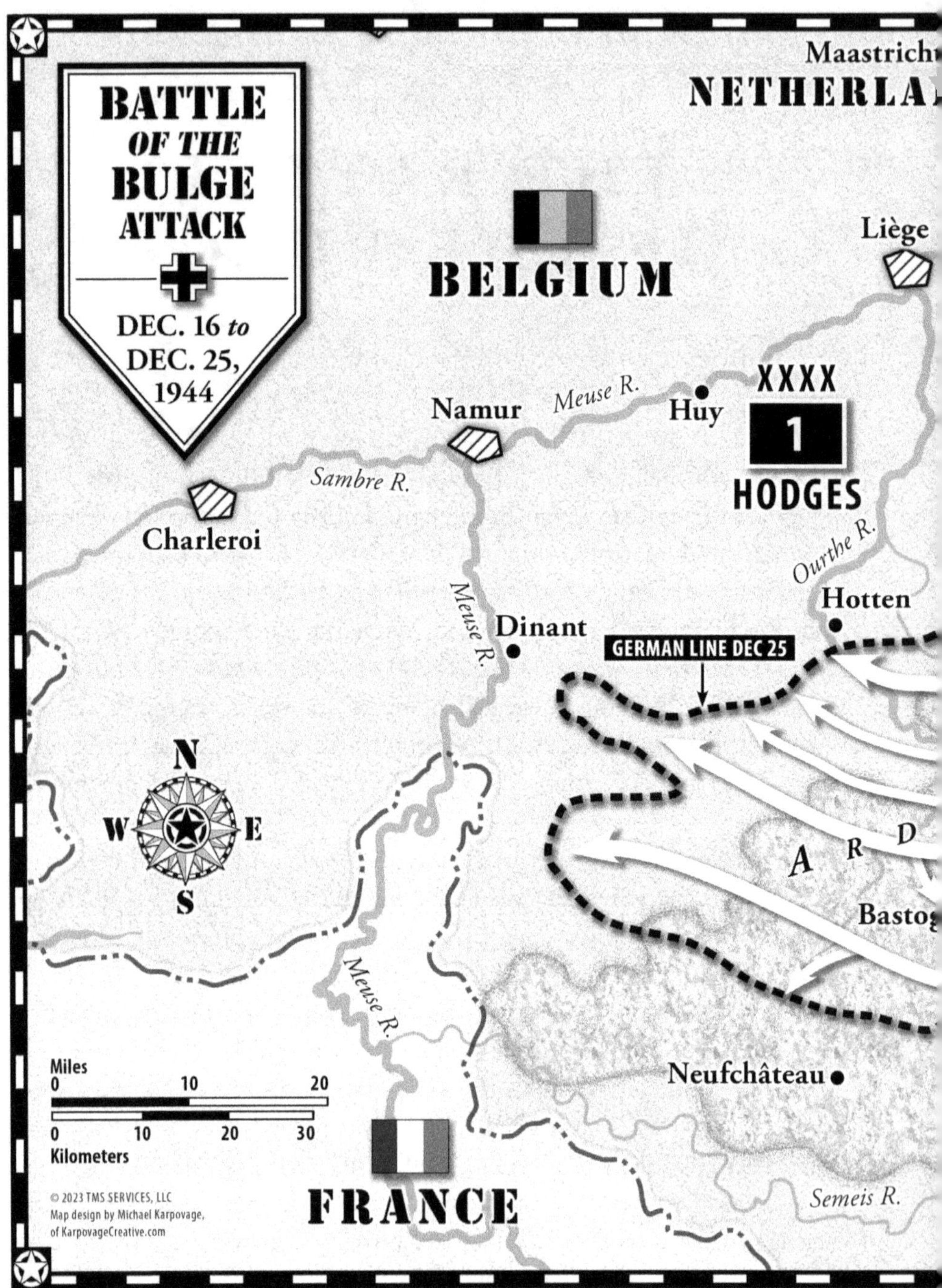

MAP 12

Aachen
Düren
Bonn
Erft R.
Rhine R.
Clermont
HÜRTGEN FOREST
Roer R.
GERMANY
Eupen
XX
Verviers
V
SEE DETAILED MAP
XX
1
WEST WALL
Mosel R.
Spa
III
26
1
XXXX
Bütgenbach
6 SS
oumont
DIETRICH
XXXX
St Vith
5
AMERICAN LINE DEC 16
MANTEUFFEL
N E S
Our R.
XXXX
7
BRANDENBURGER
Mosel R.
Diekirch
WEST WALL
Our R.
Trier
LUXEMBOURG
Arlon

Of the three attacking German Armies, the 6th Panzer Army, later known as the 6th SS Panzer Army, was the main effort. It attacked in the north, an area that became known as "the Northern Shoulder of the Bulge." This right flank of the counteroffensive was to run from Monschau, just south of Aachen, along the Albert Canal and into Antwerp. This put Clermont, Liege and the Schmetz farm right in its path.

The Respite Ruined

During their short rest period, the GIs of D Company, 26th Infantry, 1st Infantry Division, had grown quite fond of the peace, quiet and comfort of their home-away-from-home. To them, it was four-star accommodations. Conversely, Maria enjoyed pampering her "boys," Henri appreciated the extra "farmhands" and Henri, Jr. had had his own "big brothers." As for Marcel, he was in awe of those big, strong, friendly—but most of all—liberating Americans. They were his superheroes. Despite the hardships they had endured, the horrors they had witnessed and the vicious fighting they had executed, these men were gentlemen who treated the Schmetz family like royalty.

But the promise of a full month-long respite turned out to be too good to be true. In the very early hours of December 17, a jeep came roaring up to the Schmetz house, just as General Andrus had done a few short weeks earlier. This time there was a real sense of urgency. *"We're moving out,"* the driver yelled. Upon finding D Company Commander Captain Steve Phillips, the driver explained that the Germans were breaking through all along the Ardennes front. Captain Walter Stevens commanded a quick huddle with his staff, platoon leaders and sergeants. "Weapons and ammo only!" the First Sergeant exclaimed. *"We're moving out NOW!"*

"What is happening?!" thought Henri, Maria and their boys. In no time, the 82 mm mortars, heavy machine guns, all the accoutrements of a heavy weapons company and the ammunition to make them deadly were gathered, readied and loaded onto trucks. They were ready to move out. In the urgency of war, the men of D Company had to leave their personal belongings behind. Marcel would guard them as if they were treasure, assuming his heroes would one day return to collect their property.

Charles Avon

PFC Charles Avon, from Canton, OH, stood next to Maria's stove in the kitchen. Tears streamed down his cheeks. The night before, he had had a premonition this would be his last battle, and he longed to see his infant son again. After a moment, he swallowed his sobs, gathered his gear and vowed to himself he would never let his fellow troopers down. Maria gave him a long embrace. Years later, Avon's son would relate that his father was severely wounded in the ensuing battle. Though he would never completely recover from those wounds, Charles Avon was able to live a long life with the family he had feared he would never see again.

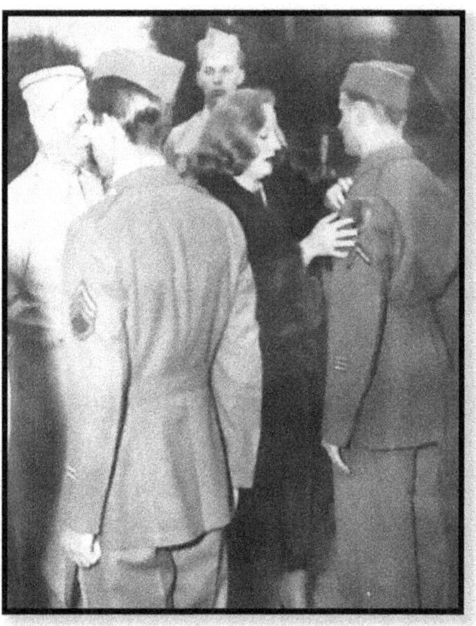

Private Charles Avon receiving the Purple Heart from actress Tallulah Bankhead.

Charlie Amero, the machine-gunner-turned-art-teacher who had shared and compared drawings with Henri, Jr., gave Henri the set of colored pencils he had carried with him since Omaha Beach.

As he was rushing out, another soldier, whose name Marcel cannot recall, approached Marcel, handing the young boy his personal wristwatch.

For Catholics in Belgium, it is tradition to get a new wristwatch for your Solemn Holy Communion (Americans call it First Communion). In 1945, Marcel proudly wore that soldier's watch to his religious coming-of-age. In addition to its immeasurable importance to Marcel, the watch turned out to have significant monetary value, too. As a young man, Marcel took the watch to a local jeweler to be repaired. The jeweler told Marcel the watch could not be repaired and offered him another watch instead. Marcel recognized the scam and still has the precious memento today.

Clayton Ramaly would later tell of little Marcel standing at attention, saluting and waving goodbye to his heroes.

And so, once again, D Company slogged off to war. A few days later, rear-echelon units came to the Schmetz farm to pick up the field kitchen and take it forward to D Company. As a thank-you for the hospitality the family had shown their brothers-in-arms, those soldiers left behind several 50-pound sacks of flour, three cases of D rations and a 20-pound can of fresh whole coffee beans. For many mornings after, Maria baked fresh bread and gave each of her boys a piece of chocolate before school and chores. In years to come, this loving ritual would again become an important part of Marcel's life.

Clermont Becomes A Fuel Depot

When the Americans liberated Belgium during the first two weeks of September, they began moving their ammunition and fuel depots further forward. This was intended to keep supply lines as short as possible. While Eupen and Spa were two of the larger supply points, even tiny Clermont became a fuel depot. Five-gallon "jerrycans" ("Wehrmacht Kanister") were stacked by the thousands in Clermont's town square. The steel cans were designed by the German army to transport 20 liters of fuel. The nickname "jerrycan" was contrived by British soldiers, using the derogatory nickname—Jerry—they used to refer to the Germans.

While the full size and severity of the German attack was not immediately clear, the potential threat to these fuel depots clearly was. The Germans' Achilles heel was a lack of fuel reserves for operations beyond only a few days of combat. Their attack plan counted on capturing American fuel, so their Panzer tanks could make it all the way to Antwerp. To keep the fuel from falling into German hands, several Army combat engineers were sent to Clermont to blow it up. Freddy Rousselles, the local liaison to the American troops in the area, met them in the town square as they were preparing charges to detonate the stockpile.

Simultaneously, Laurent Veeschkens watched from the town hall. Immediately, he became alarmed by what he saw. Running from the town hall, Veeschkens screamed, *"What are you doing? If this blows up, the town square and surrounding homes will go with it!"* A quick thinker, Veeschkens pitched an idea to the engineers. The Communal House that overlooked the town had a wedding hall on its upper level. From there, the two roads leading into Clermont from the east could be clearly monitored for a great distance. Veeschkens proposed that if there was any sign of approaching Germans, then, and only then, the charges could be detonated. The engineers checked with their superiors, the new plan was approved and Freddy Rousselles and the engineers set up shop in the wedding hall. The townspeople were happy to feed and supply them, grateful they might prevent the loss of their town center. By the third day, it became clear the German attacks were being diverted to the south and west. The 1st Infantry Division, 30th Infantry Division, 82nd Airborne and the remnants of the 99th Infantry Division had repelled the Germans with an ever-strengthening barrier. Thanks to them, the quick-thinking Laurent Veeschkens and Freddy Rousselles, Clermont was spared death by conflagration.

The 26th Infantry's Stand At Dom Bütgenbach

D Company deployed from the Schmetz farm, along with the rest of the 1st Battalion from their own rest areas. They undertook a night road march, spearheading the 1st Infantry Division's move to Camp Elsenborn, about 32 kilometers (~20 mi.) southeast of Clermont. They arrived at 0700 hours on December 17. The Big Red One planners, along with surviving local command elements, took stock of the tactical situation and determined the best defensive posture. The division was immediately redeployed south to the town of Bütgenbach.

It was an all-hands-on-deck situation as the Big Red One had nowhere near enough time to receive all the replacements and equipment they required to be at full strength. Cooks, clerks, band members and mechanics were all placed on the line. Bob Blett was still recuperating in Liege when a Captain came in with a clip board, calling names. Blett's was one of them.

DEFENSE OF DOMAINE BÜTGEN-BACH
DEC. 21, 1944
0900 to 1430
To Bütgenbach
(1 mile)
E 18
C (-) 26
2C TD 613
C 26
598m
589m
2 26
Dom Bütgenbach
G 26
MORETTO'S DASH
H 26
MANOR HOUSE AND BARNS
C 26
G 26
2 AT 26
E 26
G 26
MLR
5 1
3 1
6 1
Yards
0 250 500
0 100 200 300 400 500
Meters
BÜTGENBACHER HECK
To Morschheck
(300 meters)
613m
© 2023 TMS SERVICES, LLC
Map design by Michael Karpovage,
of KarpovageCreative.com

MAP 13

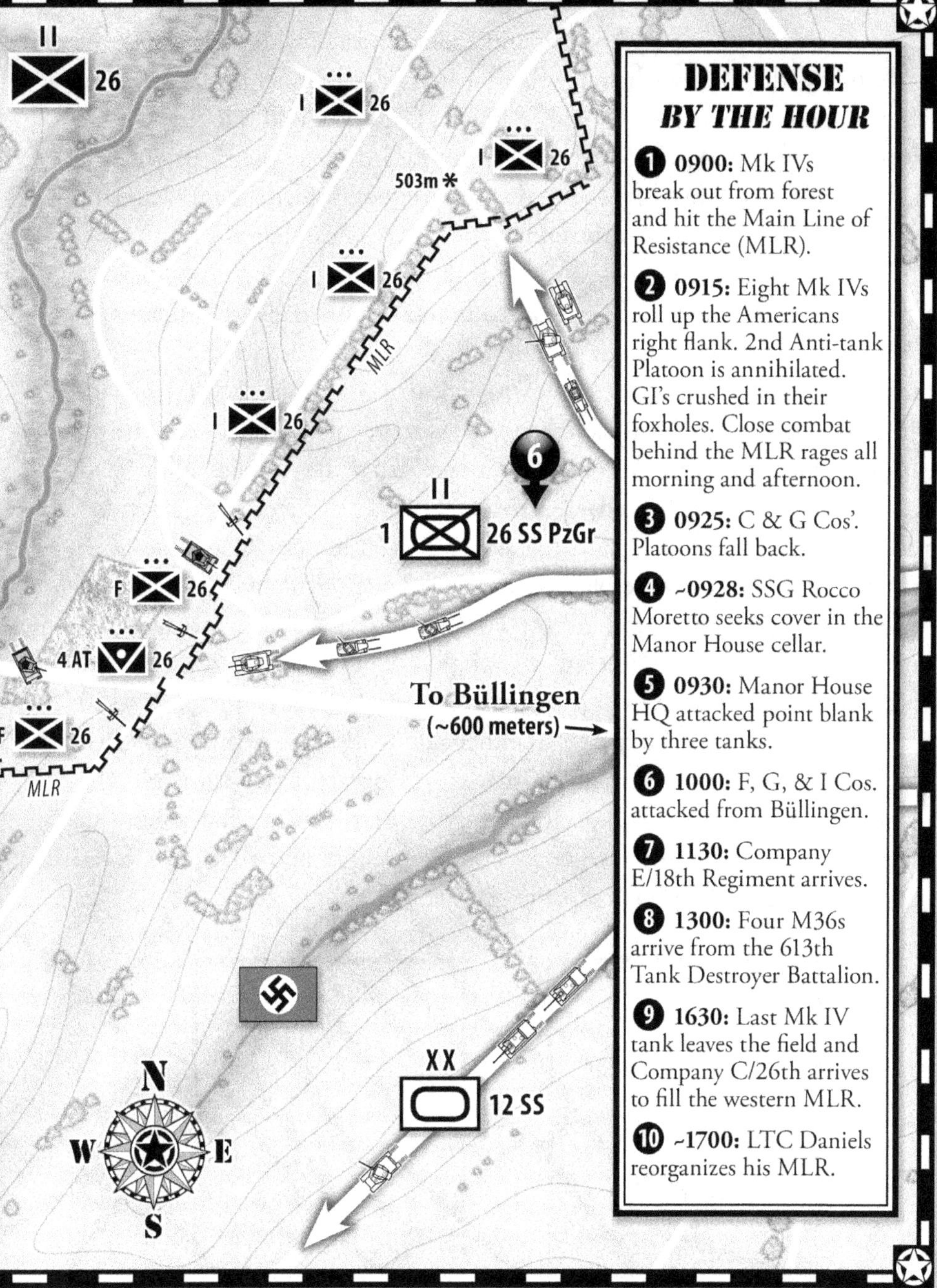

II
26
I 26
I 26
503m
I 26
MLR
I 26
6
II
1 26 SS PzGr
F 26
4 AT 26
F 26
MLR
To Büllingen
(~600 meters)
XX
12 SS
N
W
E
S
DEFENSE
BY THE HOUR
1 0900: Mk IVs break out from forest and hit the Main Line of Resistance (MLR).
2 0915: Eight Mk IVs roll up the Americans right flank. 2nd Anti-tank Platoon is annihilated. GI's crushed in their foxholes. Close combat behind the MLR rages all morning and afternoon.
3 0925: C & G Cos'. Platoons fall back.
4 ~0928: SSG Rocco Moretto seeks cover in the Manor House cellar.
5 0930: Manor House HQ attacked point blank by three tanks.
6 1000: F, G, & I Cos. attacked from Büllingen.
7 1130: Company E/18th Regiment arrives.
8 1300: Four M36s arrive from the 613th Tank Destroyer Battalion.
9 1630: Last Mk IV tank leaves the field and Company C/26th arrives to fill the western MLR.
10 ~1700: LTC Daniels reorganizes his MLR.

It had been less than a month since he was wounded; by December 18 he was back with C Company. Captain Donald Lister, who had taken over for Captain Allan Ferry when he was captured in November in the Hurtgen Forest, welcomed his medic "home." Sergeant Clayton Goode picked up Blett and told him he had a nice foxhole waiting for him, one the Germans had dug earlier. Goode was now the temporary platoon leader since Lieutenant Brooks had been killed near the end of the Hurtgen campaign.

With its three battalions, the 26th Infantry took up blocking positions a few kilometers south of Bütgenbach. It was a crossroads, dominated by a large manor house, known as Dom Bütgenbach. Just to the south of the Dom was the Morscheck Crossroads, a major route to Bütgenbach, where precious American fuel and supplies had been stockpiled. Capturing that stockpile would give the 6th SS Panzer Army the supplies Hitler had promised, but not delivered. The Morscheck Crossroads also provided a major route for the Germans to push west. As it turned out, the 26th Infantry Blue Spaders were placed directly on one of the five major attack routes of the 6th SS Panzer Army. For the next five weeks, their entire universe became a one-kilometer arc to the south of the manor house and its outbuildings. For the individual infantryman it was claustrophobic—100 meters to his front and 50 meters to either side—all while living in a frozen hole in the ground with a make-shift roof to stop artillery fragments. The Hürtgen Forest left the 26th decimated, particularly its 2nd Battalion. On December 17, 2nd Battalion was only at 80% strength and 60% of that was new replacements with no combat experience.

Dom Bütgenbach in 1948 and in 2014.

The 2nd Battalion bore the brunt of the German drive to Bütgenbach, with the 1st Battalion supporting its west flank. C Company, fresh from its own rest in Aubel, was soon attached to the 2nd Battalion. They were holding a defensive position between Dom Bütgenbach and the Morscheck Crossroads. D Company, part of 1st Battalion, had some of its heavy weapons distributed across the C Company element of the 2nd Battalion. This was the 26th Regiment commander and his three battalion commanders making ad hoc adjustments to deploy their forces to improve the chance of success.

PFC John Lebda, a machine gunner from D Company when he landed on Omaha Beach, was now in charge of the D Company motor pool and its vehicles that were used to haul the heavy weapons and ammunition. Ongoing losses of men forced this job change, as Lebda was qualified to drive and maintain both wheeled and tracked vehicles. The heavy, frequent German attacks generated an urgent need for mortar shells for D Company's 81-mm mortars, and mortar ammunition consumption was higher than at any time before. The heavy snow prevented the use of wheeled vehicles, so Lebda was ordered to use the one tracked vehicle—called a "Weasel"—their battalion had. In addition to numerous trips to his mortarmen, Lebda hauled ammunition, mines and barbed wire to his C Company riflemen deployed forward with the 2nd Battalion. He was a godsend to Rocky, Sam and Bennie.

The 12th SS Panzer Division, 3rd Fallschirmjäger Division (the paratrooper branch of the German Luftwaffe) and the 12th Volksgrenadier Division were all arrayed against the 26th Infantry Regiment, 1st Infantry Division. The 2nd Battalion commander, Lieutenant Colonel Derrill M. Daniel, knew what was at stake. He told his officers to tell their men, *"We fight and die here."*

On December 21st, a fierce German attack by the 12th SS Panzer Division (Hitlerjugend) overran the western edge of the 26th Infantry's line. Rocky and his platoon fell piecemeal back to the large, sturdy Manor House. The mighty American artillery killed many of the German infantry accompanying the tanks, but ultimately three Mk IV panzers made it to the barn only meters short of the Manor House. The three tanks with their 75mm guns pulled up tight to the south side of the south barn on the property. In that position they could fire point blank on the Manor House as well as the building to the northwest of the house where E and

H Companies had their headquarters. The 2nd Battalion commander, Lieutenant Colonel Derril Daniels, had his headquarters in the basement of the Manor House. Moretto, and several other soldiers made a beeline for the basement. Infantrymen learn early on that cellars are the safest place when in doubt. There Moretto and several other retreating soldiers joined LTC Daniels and his staff at the 2nd Battalion headquarters.

Rocky gives a first-hand account: "Colonel Daniels was personally directing artillery fire over the radio. He was in communication with all sorts of artillery units including our own 33rd Field Artillery Battalion and the Divisions 5th Field Artillery Battalion with their 155's (155 mm artillery guns). He even was asking for corps artillery and at one point he yelled over the radio, *"Get me all the damned artillery you can get."* There is no doubt in my mind that Colonel Daniels almost single-handedly slowed the German advance until reinforcements slowly began to arrive and started to build up our positions.

> ***"Thanks to Colonel Daniels and fortunately for us the German Infantry had taken all sorts of casualties from the artillery fire and were unable to penetrate our defenses in any number. When Colonel Daniels was advised about the two tanks which had penetrated to within 20 yards of the building, he asked to be kept advised of their movements. I would inch up the cellar stairs and when the tank crews would spot me they would turn the 88s and fire a round. But before they did, I would come flying down those cellar steps.***
>
> ***"The situation remained that way it seemed for an eternity. Colonel Daniels continued with the artillery fire and then called for fire directly on our own positions in an attempt to drive the enemy off us. He then called for volunteers to knock out the tanks with a bazooka. One young soldier somehow with help managed to get on the roof of the farmhouse and miraculously disabled one tank. It seemed like an impossible task but***

> ***somehow that kid got the job done. The remaining tank stayed for a while and then turned tail, probably realizing he was sticking out like a sore thumb without support. It was fortunate for us that our artillery inflicted so much damage to the German Infantry, otherwise we would have surely been outflanked."***

The 1st Infantry Division fiercely defended their line for four more weeks until adequate U.S. forces were in place to "reduce the bulge". In the fourth week of January 1945, the battered 2nd Battalion, 26th Infantry was replaced by the 1st Battalion. C Company was given the order to attack and take the Morscheck Crossroads. In less than a day, C Company took its objective, and successfully repelled a counterattack. When all was said and done, the Germans never made it through Bütgenbach, from either the Morscheck Crossroads or Bullingen. The Americans had held their line at great cost to both sides, though the German losses were much greater in men and armor and irreplaceable.

Sergeant Rocky Moretto, the kid from Hell's Kitchen in Brooklyn, NY, was one of the C Company attackers who took the Morscheck Crossroads. A few weeks earlier, while D Company was bivouacking at the Schmetz farm, C Company was bivouacked in Mathilde's hometown of Aubel. Just like all the Big Red One men resting in Liege Province, passes were given to those who wanted to travel. Paris was an option, but much of the pass would be spent on the road. Spa, a resort town known for its healing mineral springs, was another destination. This is where Combat Engineer Bill Clements chose to relax. Verviers was yet another destination. At Verviers, there were USO shows, movie theaters and hot showers. Not only was it closest to Aubel, but it was also a town that C Company had liberated on September 11.

Moretto and his best friend, Tech Sergeant Bob Wright, from Michigan, and the rest of C Company, had been enjoying R&R not far from the Schmetz farm near Aubel. They took advantage of the Verviers option so they could surprise the Lemaire family with a visit. Good food, comfortable beds and young attractive women may have had something to do with their decision. There, Wright would reunite with Lucienne Lemaire. Rocky said it was the happiest he had seen Bob since they had left the States.

Bob Wright with the Lemaire family.
(l. to r.) Wright, Josette, Lucienne and Rene Lemaire

A few days later, as the two friends were digging in together to hold the Crossroads they had captured, an enemy artillery shell landed just outside Moretto's foxhole. Bob Wright was killed instantly. The shrapnel missed Rocky completely.

Sergeant Bob Wright (l.) and Staff Sergeant Rocky Moretto (r.)
near Dom Bütgenbach the day before Wright was killed.

Robert Wright was laid to rest at Henri Chappelle Military Cemetery, just up the hill from the Schmetz farm. Rocky believed that Bob would have found Lucienne after the war and lived happily ever after. As it was, Lucienne married another American soldier and moved to New York after the war. She saw Rocky once more, this time for coffee in New York City at Rocky's workplace. Rocky Moretto spent 11 months in combat, fighting from Omaha Beach to the end of the war. He was never wounded or captured. All the men who fought with Rocky knew him to be an excellent soldier, but Rocky was the first to admit that luck was just as important as soldiering skills.

By this time, the suffering and fear endured during occupation and annexation had been extinguished for more than four months. For many Belgians, the war was over and had finally moved on to the enemy's homeland. But not for those in the "bulge" created by the German counterattack. Their nightmare would last another six weeks. Fortunately, the Bulge would never quite extend far enough to the north to affect Thimister-Clermont, in large part thanks to the Big Red One's stand at Bütgenbach.

Some of the most vicious fighting in Europe occurred during the six weeks from December 16, 1944 to January 30, 1945. At least 19,000 Americans were killed during the Battle of the Bulge. Another 23,000 went missing. Many were captured. Some eventually caught up with their units. But there was still a lot of fighting—and dying—to be done. The men on both sides needed a rest. Once again, American units fell back into Belgium for some respite. This time, it was units of the 99th Infantry Division who bivouacked in Clermont.

Tending To The Fallen

During this time, Marcel had another experience that would forever cement his love for, and dedication to, the American soldiers. Staff Sergeant Vito Mastrangelo and his men continued their solemn pursuit of collecting, preparing and burying the thousands of fallen men, less than three kilometers (1.9 mi.) from Marcel's home and less than 0.5 kilometer (500 yds.) from his school. Henri Jr. and Marcel knew there was something important going on just across the National Road 612 (N612), next to the Putters' farm. Whatever it was, it was fenced off from the public and guarded by soldiers. The boys assumed it was a cemetery because they would regularly see open trucks filled with bodies. Some were headless, some were missing limbs and most of the trucks spilled blood onto the road as they went by. Marcel recalls seeing more than 200 bodies each time he could get close enough to observe the heavy traffic. The Schmetz brothers saw, first-hand, the macabre results of the Battle of the Bulge. For Marcel, this was far more soul-reaching than his experiences with the healthy young men he had played with on the farm just weeks before. Nightmares followed, but so did a quiet, but deep, resolve to do something. Somehow, Marcel would help repay this incredible sacrifice. ✪

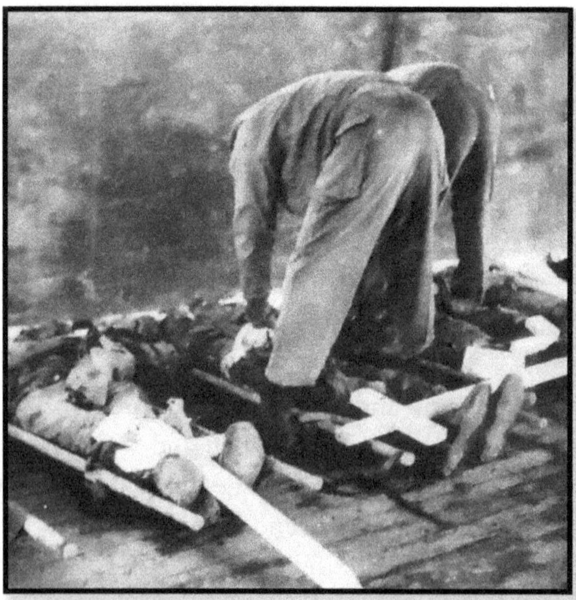

The 607th Graves Registration Company at Henri Chapelle Cemetery.

8

POST-WAR

MOST OF BELGIUM WAS LIBERATED IN SEPTEMBER 1944. But it was not until February 1945 that the entire country was free of the Wehrmacht. Freedom and liberty reigned once again. But food was still scarce, and many Belgians were malnourished. Infrastructure had been damaged, destroyed or transferred to Germany. Of the more than 81,000 Belgian dead, 86% were civilians.

Then, there was the question of what to do with collaborators. It has been estimated that roughly 53,000 Belgians were found guilty of collaboration. There were those who fought in German units, including the infamous SS. But the majority were people like Gerard Krafft, who were active only at the local level and who had not collaborated voluntarily. Some, like Krafft, risked their lives by covertly breaking the German laws. Nevertheless, the re-established Belgian bureaucracy came for Gerard, and he was arrested. His friends and neighbors—especially Henri Schmetz, Sr.—testified that Krafft had gone to great lengths—at great risk to himself and his family—to shield Thimister-Clermont from the worst of German policies. And, he had actually conspired with Henri to aid escaped prisoners-of-war and forced laborers. Despite the testimonies, Gerard Krafft was sentenced to prison in nearby Verviers. Ongoing appeals failed and he served one year before being released. Marcel remembers visiting him and bringing along his mother's homemade soups and baked goods.

A Fortuitous Accident

It was a cold, clear day as Mathilde Meeuwissen drove her extremely old Mercedes-Benz along N612 from Aubel to Herve to do some shopping and run some errands. Her 15-year-old son, Marc-Antoine, rode in the front seat next to her. He enjoyed riding "shotgun," because he had a better view of the rolling farms and countryside, of which he never tired. As Mathilde entered the Henri-Chapelle American Military Cemetery, she paid little attention to the 8,000 American soldiers to her left. After all, they had been there for almost 50 years, and there was some type of organized memorial event—just for them—at least twice a year. For many Belgians, that was the extent of their knowledge of the unfortunate young men who came to be buried so far away from their homes and families.

Born after the war, Mathilde was the first child of Joseph and Guillemine Meeuwissen. Joseph was born and raised in the Netherlands. During the war, he was captured by the Germans and transported to work as a forced laborer on a farm in Sippenaeken, near the Dutch-Belgium border. Later, he was sent to work on another farm near Aubel. While at the second farm, he met Guillemine, who was the daughter of a neighboring farmer. Of course, the Germans, and Joseph's situation, made it next to impossible for the young couple to be a couple.

After the war, Joseph relocated to Aubel, where he had an opportunity to farm for himself. It also gave him the opportunity to rekindle the relationship with Guillemine, and soon they married. The occupation of Aubel, and Joseph's forced labor for the Reich, left bitter memories for the young couple. Food was available but not like it was pre-war, and it took some time to reverse the chronic malnourishment. Because the Meeuwissens did not want their children to be affected by their own troubles and sufferings, the war was never discussed in their home. And it was only marginally covered in school. As a result, for Mathilde, it was not that she did not care; she just did not know to care.

Growing up, Mathilde was the oldest of nine siblings. She excelled at school and enjoyed studying Latin and Greek. She dreamed of someday becoming a doctor and helping people. But her dream would need to be put on hold.

The Meeuwissen Family in 1972. Front row: Agnes, Guillemine, Gerard, Brigitte, Joseph. Back row: Marie-Josee, Antoine, Marie-Jeanne, Arlette, Mathilde, Pierre.

After giving birth to several children, Guillemine began suffering from a debilitating medical condition. By the age of 36, she was no longer able to function as a mother or housewife. It was never clear if Guillemine was overcome by post-partum depression, chronic fatigue syndrome or another chronic malady. But, Joseph pulled Mathilde from school and soon her life became cleaning house, feeding babies, making meals and dressing young ones for school. She also had to help in the fields and milk the cows, with the help of her grandfather. Before she knew it, all the rites-of-passage and experiences of a young girl's teenage years had passed her by. Any hopes of attending medical school had been dashed. Mathilde loved her parents and siblings, but her life was smothering her. Her plan for self-preservation was to get married, move out and start a family of her own. As soon as possible.

Mathilde Begins Her Own Life

Soon after graduation, Mathilde met a man several years her senior. He was a complex character, but showed her affection, something she felt was lacking in her life. Things progressed quickly and the man proposed. Joseph was concerned about what Mathilde might be getting herself into. He had seen concerning behaviors by his prospective son-in-law, and Joseph was very much against the marriage. But for Mathilde, the hope of a better—or at least different—life was too strong. The couple married in 1967. Soon Mathilde was expecting her first child. At the same time, to Mathilde's amazement, Guillemine was expecting her ninth child! There would be 20 years between Mathilde and her youngest sister. Mathilde knew she would again be changing diapers and making bottles, but at least it would be for her baby, not her mother's. A daughter, Francoise, was born the following year.

Mathilde would have two more children. Benoit was born in 1969, Marc-Antoine in 1975. Benoit, the middle child, seemed to develop differently than his brother and sister. Eventually, he was diagnosed with a mental-health condition that would occasionally require him to be institutionalized. This put immense stress on the marriage, especially for Mathilde's husband. He suffered mood swings that started mildly, but became more frequent and more intense. Ultimately, he committed suicide, an outcome his family believes was the result of bipolar disorder.

Suddenly, Mathilde was a stay-at-home mother with three young children—one with special needs—and no source of income. To generate a meager income, Mathilde took work cooking for large families and groups.

It was what she was best prepared to do well. It was also an ironic twist of fate as she resumed the life she had fled when she got married.

Mathilde's Fortunes Take A Turn

As Mathilde passed between the golden columns on either side of N-612, the virtual gateway to the Henri-Chapelle American Military Cemetery, she wondered what the future might bring. Deep in thought, she hoped for better luck and more happiness. Suddenly, a snow squall appeared. A car, coming around a turn, approached her in the opposite lane of the two-lane road. The car hit a patch of black ice, which sent it skidding into Mathilde's lane. The vehicles collided, the skidding car striking Mathilde's vehicle almost head-on. As her car recoiled, spun and came to a sudden stop off the road, she reached over to check her son. He seemed OK, though a clump of his hair was stuck in the starburst crack in the windshield in front of him. Cemetery workers ran to the vehicles. The workers pulled her car from the ditch, off the road and into the nearby utility area of the cemetery.

N602, Mathilde's accident location at Henri-Chapelle Cemetery.

Mathilde was crying. Though not severely injured, she was shaken. But mostly she was upset that her car appeared to be "totaled." She desperately needed it to work and earn a paycheck. And she realized a new car would not be in her budget any time soon. The cemetery staff were wonderful. They drove Mathilde and Marc-Antoine home. On the ride home, she wondered how she would pay her bills and feed her children.

Mathilde needed to get her car fixed, and none too soon. But how would she find an honest, trustworthy body shop owner? This was something her late husband used to handle. She went through the phone book and found a body shop in Clermont, not too far away. That was it. She would put her faith and hope in someone named Marcel Schmetz.

But Marcel was always very busy. After taking her initial call, it took him three days to pick up Mathilde in Aubel, and drive to the cemetery from where her car had to be towed. Though still nervous and cautious, Mathilde thought to herself that a man who was always that busy must surely be reliable and do great work.

Henri, Jr. Finds A Bride

After the war, both Marcel and Henri, Jr.—now a young man—continued to work on the family farm. But Henri longed to get married and start a family of his own. As was obvious from his many drawings and paintings, Henri loved animals and wanted his own farm and livestock. But, Henri was never again the same extroverted teenager he was before spending 15 months, hiding from the Germans and the rest of the world. He was not even the same outgoing lad who had mingled so well with the D Company soldiers who had stayed on the farm. The time spent in isolation had taken a heavy toll on Henri. Marcel believes it was survivor's guilt from not having done his duty to fight or resist the Nazis.

Eventually, Henri found love and married a local girl named Laure. They leased a farm in the nearby town of Charneaux and purchased dairy cows. His new wife, and the responsibility of the new farm, seemed to be the tonic Henri needed to improve his outlook. Before long, the young couple had a daughter they named Maria, in honor of Henri's loving, hard-working mother.

Henri, Jr. and Marcel in May 1945.

Henri, Jr. with wife Laure and Maria at Maria's First Communion.

Henri, Jr. was on a good trajectory, but it seems that happiness was not to be for him. Laure was riding her bicycle to the local market to buy a few staples for her family. She was struck and killed by a motorist not far from her home. Henri, Jr. was left to raise a young girl and tend to his small farm by himself. He became withdrawn and even more taciturn. Like Mathilde Meeuwissen, young Maria would come to learn precious little about her father's wartime experience.

Marcel Comes Into His Own

Marcel, for his part, liked to tinker with things. He would take things apart, learn how they worked and put them back together. He also had a passion for building things from scratch. All kinds of things. It soon became obvious that farming was not going to scratch that itch.

Not far from the farm on Quoidbach was an auto repair shop. Though he had no experience, one day, Marcel approached the owner, and convinced him to bring Marcel on as a mechanic. Over the next five-year "apprenticeship," Marcel not only learned the auto mechanic trade, he also mastered collision repair. And, as it turned out, he was very, very good at it.

By 1965, Marcel had saved enough of his hard-earned money to purchase a 300-year-old house with a few small outbuildings. The house was barely two kilometers (~1.2 mi.) from his parent's farm. Marcel tore down the smaller structures and began to construct his new life and livelihood. He built a garage where he could practice his craft, and soon became the go-to-guy across Thimister-Clermont.

After Marcel had settled into his new home and established his business, he slowly began to move his precious war memorabilia from the farm to the back of the garage where he kept his personal vehicle. The personal effects of his American heroes—left behind on the farm decades ago—and his soldier's watch were Marcel's most precious belongings.

Belgium, and most of Europe, had effectively became a "landfill" for the detritus of war. Weapons, ammunition (spent and live), mess kits, helmets, thousands of kilometers of field telephone line and a myriad of war materiel laid where it had fallen. For decades (and even to this day) farmers have found relics in their fields while plowing after the spring thaw, as did hikers walking trails in the woods. On occasion a soldier's remains would be found. People around the area were aware of Marcel's interest in war materiel, and would give their findings to him or use them for barter for work on their vehicle.

It was hard to put into words how much these items meant to the man who, as a boy, had lived with 110 American soldiers, even if only for a brief time. Marcel owed his freedom and his life to those young men and their comrades. He thought of them often, especially when he glanced upon his collection. Those soldiers had spent only a few weeks with Marcel and his family, but they had left an indelible impression.

Marcel's Family

Marcel's father, Henri, Sr., had a good life. He survived two invasions and occupations, served his country and raised a family with his beautiful wife. His family and his farm completed him. Though he was a prisoner-of-war, that miserable time taught him important life lessons. When Henri, Jr. and Marcel left the farm to pursue their own life interests, Henri, Sr. realized he could not keep up with the farm by himself. In 1967, he and Maria retired and went to live with Marcel in his new home. The businessman from Liege who owned the Schmetz farm was very sad. Knowing he would never find such wonderful tenants again, the businessman sold the property. Charles Putters, who as a young boy in 1944 had watched the Graves Registration units bury soldiers next to his father's farm, eventually bought the house. He and his wife still live there today.

Henri Schmetz, Sr. passed away in 1984. He is buried next to the church in Clermont. Maria stayed with Marcel for her remaining years. She would chat with customers when they dropped off or picked up their cars. Or have coffee with those who waited. She became the best advertising the shop could ever have. When Maria passed away in 1991, Marcel was by himself. But he was happy, as most of his customers were, or would eventually become, his friends.

The Consummate Craftsman

Marcel was exceptionally good at his job. Many considered him more of an artist or craftsman. A magician, rather than a body shop guy. Marcel loved his work. He also loved his home, putting endless effort into modernizing the old building. He was comfortable with his life, and the way things were. He considered himself an affirmed bachelor; a wife and children held no appeal whatsoever. As he told any friend who would inquire, *"No wife, no kids, no trouble!"*

One day, a lady he did not know called the shop to ask for help. She said her name was Mathilde. His schedule was so full he had to make her wait three days before he could deal with her problem. Her car was at the Henri-Chapelle cemetery, so he picked her up in Aubel, and they drove together to see the damaged vehicle. It was an old Mercedes-Benz with significant collision damage. Marcel had completed complicated, challenging repairs before, but this car was so old he knew that finding replacement parts was going to be a problem.

After working on the wreck for several days, Marcel concluded that the best solution would probably be to sell the remaining good parts of the car. He presented the idea to Mathilde, who reluctantly agreed. Marcel set about disassembling the car and selling its parts for as much as he could negotiate. Whenever he accumulated a significant sum, he took the money to Mathilde. She was becoming despondent. Public transit was not available in the rural areas, and her jobs were just too far to bicycle, even in good weather. She simply could not do her job without transportation. *"I need a car as soon as possible so I can earn money to pay my bills,"* she said, tearing up. Marcel felt terrible. All he could muster was, *"I will do my best."*

Mathilde began to think that maybe a little well-intentioned bribery might expedite things. To try to keep her car his top priority, she began to make soups, bringing generous amounts to Marcel whenever she stopped by. And she stopped by often. Marcel was amazed at how good the soups were. In some cases, they were even better than the soups his mother used to make. After a couple of weeks of culinary encouragement, Mathilde still had no car. So she upped the ante, made a complete dinner and took it to the garage. Maybe Marcel would work later, she thought, if he knew a full dinner would be ready for him. The strategy paid off. Finally, together they found a second-hand car for Mathilde. The cannibalization of her old Mercedes contributed greatly to the purchase.

The Relationship Deepens

Mathilde came to realize that Marcel was a kind, industrious, sincere man. He seemed genuinely interested in her, and always asked about her children. And, he was ever-so-apologetic about how long it was taking to finish her car. He was, however, 15 years her senior. Marcel, of course, had noticed immediately that Mathilde was a young and quite attractive woman. She was a hard worker, who doted on her children. And he really liked the delicious food she brought him.

It was inevitable. Age aside, they made a perfect couple. Each filled a void for the other; each completed the other. On June 6, 1992—the 48th anniversary of D-Day—Mathilde and Marcel were married. Marcel adopted all three of Mathilde's children and treated them as his own. ✪

Mathilde and Marcel on their wedding day, June 6, 1992.

9

GRATITUDE IS NOT ENOUGH

TO SAVE MONEY BEFORE THE WEDDING, MATHILDE had moved in with Marcel a few months earlier. Naturally, being a wonderful wife and housekeeper, she began cleaning, organizing and putting a woman's touch to her new home. One day, she ventured into the garage, her fiancé's man-cave. She planned to surprise Marcel by cleaning and organizing the space.

Rummaging inside the dusty garage, she found Marcel's car, his tools and equipment, some old personal items and a whole lot of parts for who-knows-what. Delving a bit deeper, she discovered a pile of green duffel bags that appeared to belong to Americans. Along with the duffels was a cache of old military-looking items.

She pondered the intriguing findings for a few days. Then, one evening after Marcel had finished with work, Mathilde quietly asked him about the strange items she had found in the garage. Marcel took several long, deep breaths. Then he told her the most awe-inspiring story she had ever heard.

A Life-changing Story

It began with the invasion of Belgium by the Germans in 1940, some eight years before Mathilde was born. Marcel spoke of the German occupation, his father's capture and forced labor and the eventual annexation of his family's farm. He spoke of the Nazi oppression, the pervasive fear and the endless deprivation over four long years. He spoke of his older brother having to hide from everyone for more than 15 months. He told of a Company of young American soldiers—boys, really—who had come out of nowhere and asked to stay on the farm. And he recalled how—like the flipping of a switch—the family's suffering and fear had gone away.

Marcel told her how the soldiers, just as suddenly as they had arrived, were called away to fight the Germans at the Belgium-Germany border. He explained how, over six weeks, they had died by the thousands—in that so-called "Bulge"—to keep Marcel and Mathilde's parents safe and free. And he sadly recalled how he and his brother saw many of those dead piled on top of each other in large trucks heading to what is now the cemetery where Mathilde had her accident. The accident that brought them together. Finally, he told her the items she had discovered in the garage were the personal belongings those boys had left behind on the farm when they were abruptly called to battle.

Now Mathilde quietly understood that those were the same boys—7,992 of them—under the white crosses and Stars of David she had ridden by so many times on N-612. The intensely powerful story was one Mathilde never saw coming. And it was a story that would have an enormous and emotional impact on the rest of her life.

Rekindling A Fire

Many of Marcel's friends knew of his treasured memorabilia. Four in particular felt it should be put on display as a reminder of the sacrifices the Americans had made for the Belgian people. Marcel agreed. But he was always so busy, and in such demand, he could never find the time, space or energy to put it all together. Now, nearing retirement, he had a fiercely encouraging partner who rekindled a fire that had been smoldering in him since 1944. Marcel knew it was time.

The biggest question was: Where could all these unique mementos go? It became obvious the largest of Marcel's shop buildings would be the ideal place. Inspired by Mathilde and encouraged by his long-time friends, Marcel began methodically navigating toward retirement.

His first step was a decision to no longer take on new customers. This meant the large number of vehicles usually crammed in the shop slowly began to dwindle. Marcel's operation also had a smaller building, contiguous with his house. He had been using a room in the back as his auto-painting room. As Marcel's imagination raced, a plan took shape in the craftsman's head. He knew this room could be repurposed to house part of the display, and his life-long dream began to crystallize. It was now within reach.

One day in early 1992, there was a knock on the door. A gentleman in his 60s introduced himself in English. Kenneth Stanger, who had fought with the 99th Infantry Division during the Battle of the Bulge, explained that he had stayed in this very house during his R&R after the Battle of the Bulge in 1945. Stanger had slept on the second floor of the home. While at the home he had showered and received a new uniform for the first time in many weeks. This happy respite was one of many memories that prompted Stanger to seek out specific locations on his visit to his old battlefields. He had a fond recollection of his brief time at an oasis called Clermont.

The Schmetzes treated their unexpected caller like a king. After all, Ken Stanger embodied the American soldier who was soon to be memorialized by their efforts. He slept soundly in the room he had slept in almost 50 years earlier. And Mathilde made him some of the same meals she had "hooked" Marcel with not long before. Stanger wanted to repay their hospitality and generosity. Learning their wedding was scheduled for June, he invited them to honeymoon at his home in Ogden, UT. There he would show them a part of America that Europeans generally do not get to see. And so it became an integral part of the wedding planning. This was the first of many visits by the couple to the United States to visit World War II veterans.

Marcel officially retired in early 1993, quickly shifting gears to his commitment to making what he always thought was impossible, possible. First, it was determined that the large shop building that would house the main display needed a new roof. Marcel's friends got together, agreed to all take their 30-day summer vacation at the same time and the five men spent it installing a new roof on the old shop.

Marcel wanted both present and future generations to understand the hardships and suffering that came with enemy occupation, annexation and the loss of liberty and freedom. He wanted people to truly comprehend the sacrifice that had been made by a people an ocean away to restore that liberty and freedom for a people they did not even know. All at the cost of tens of thousands of their own young soldiers. But mostly, Marcel wanted people to understand the indelible impression that had been left on a young boy and his family by a hundred or so of those soldiers. He wanted all to truly "know" the men of D Company, young boys who had treated the Schmetz family like royalty, given them gifts and left behind food and supplies that frigid December of 1944.

Marcel and Mathilde gave great thought to what they would name their museum, desiring to set it apart from so many other museums across Europe that commemorated World War II. They did not want anyone to forget the privation, suffering and sacrifice of both civilians and soldiers during that time. In the end, they chose Remember Museum 39-45.

The Dream Comes To Be

The year was 1993. The 50th anniversary of D-Day, liberation and the Battle of the Bulge was a year away. Marcel and Mathilde planned to have their display ready for those special occasions. They targeted June 6, 1994 for the grand opening. It would be the 50th anniversary of D-Day. And their second wedding anniversary.

Cleaning, painting and renovating the buildings consumed them over the next months. The GI's personal articles were cleaned and sorted, with attribution to their original owners whenever possible. An even more ambitious undertaking was the creation of life-size dioramas of scenes replayed in Marcel's mind over the decades. Where possible, he used real weapons, uniforms and family belongings. He fabricated—from scratch—realistic mannequins, landscapes and even full-scale armored vehicles. The detail and accuracy fools most visitors if it cannot be touched. Many of the mannequins are based on wartime photographs, so the likeness is even more striking.

As the big date approached, Marcel learned through another friend—who had also been deeply touched by the American soldiers during the war—that 1st Infantry Division veterans would be coming to Normandy for the anniversary celebration. Afterwards, he had heard they would travel to Liege to visit their Battle of the Bulge battlefields.

Intrigued and determined, Marcel set out to see if he could identify any of the scheduled D Company veterans. His initial efforts were unsuccessful, but he did learn there would also be soldiers from C Company, 26th Infantry Regiment attending. These men had fought side-by-side with D Company throughout the European campaign and had also rested near Clermont after the Hürtgen Forest. In fact, they camped in and around Mathilde's hometown of Aubel. Among them would be Bennie Zuskin and Sam Messina—one of the C Company BAR teams—and 16 other veterans.

Marcel was able to track down Bennie and Sam, who contacted the others. After a bit of logistical maneuvering, the veterans rearranged their itinerary to come visit the new museum. They would become the centerpiece of the new official grand opening on June 12. It was a bit later than planned, but far better than Marcel and Mathilde could have ever imagined.

Those 18 American veterans became the first recipients of a long-awaited showering of love for the men who had made such an impact on an 11-year-old boy. Even then, his awestruck new wife was still comprehending the level of sacrifice these men—and their comrades who never made it home—had made for her and her family. Although the veterans could not speak French, and Mathilde spoke very little English, feelings and expressions transcended language.

The people of Clermont jammed into the town square at the gate of St. John's Church to celebrate the arrival of "their" men and the opening of the museum. Father Peters, the Catholic priest, had a special service for the men in the church. He was alive during the war and, like most Belgians, had a great respect for these men and their comrades. He was so excited that Mathilde remembers he could hardly speak. Mathilde was the organist for St. John's and was integral to the choir. The choir members spoke little to no English, but they worked long and hard to learn their favorite hymns in English so the soldiers could sing along with them. They were especially proud of the grand finale: the Star Spangled Banner.

Museum Grand Opening Celebration at St. John's Church.
l. to r. in the white Big Red One hats: Bennie Zuskin; Russell Werme; Harry Wagner (with camera in rear); John Gembel; Victor Muggar (in sunglasses); Bob Diller; Alden Peckham; Sam Messina (khaki garrison cap); Jack Hathaway; Fred Smith; a Belgian veteran in the foreground, and Marcel (white coat) and Mathilde (far right).

Before long, Bennie invited the Schmetzes to the United States for the next Blue Spaders reunion. Bennie told Mathilde he would be her English teacher when they visited. The trip came to fruition, and Bennie and his wife, Maxine, took the Schmetzes to lunch. Bennie was a man of few words and a dry sense of humor. He asked Mathilde to look over the menu and suggested that the sandwiches were very good, which she understood. He then asked her if she would like a knuckle sandwich. Wanting to be polite, she said, *"Yes,"* not having a clue what she had agreed to order. All had a good laugh when Bennie explained.

The Memorabilia

Late in World War II, the Germans introduced the V-1 "Buzz Bomb." It is considered an early version of a cruise missile. It was a frightening weapon, first deployed for the terror bombing of London, and later used to destroy or damage many Belgian targets. Cruising at 400 mph at a low altitude, the missile made a terrifying pulsating sound that could be heard from 16 kilometers (10 mi.) away. From October 1944 to March 1945, the Germans fired thousands of Buzz Bombs at Antwerp, Brussels and Liege. Antwerp alone was struck by nearly 2,500 V-1s. Many of the missiles were shot down, and many more fell short of their intended target. One such bomb landed near the Schmetz farm undetonated and young Marcel was able to observe the materials and shapes of the weapon from a distance. Years later, he was able to obtain a set of original V-1 blueprints from a German army veteran.

For the big grand opening, Marcel created an exact, full-scale reproduction of the V-1, built of plywood and detailed meticulously to the original specifications. A Belgian Army colonel gave him an audio recording of the flight, free-fall and detonation of the flying bomb. Today that replica hangs menacingly from the museum's ceiling, and the audio recording provides a terrifying reminder of the fear the 80-year-old weapon had instilled in those on the receiving end.

The Buzz Bomb replica, hand-built by Marcel.

As friends and relatives learned what Marcel was doing, they donated—and urged others to donate—additional wartime artifacts. Henri, Jr. donated an American jeep and trailer he had acquired and used on his farm. One of Maria Schmetz's most vivid memories of that time in December 1944 had been the field kitchen used by D Company outside the front door of the Schmetz home. When the Americans hurriedly departed to stop the German attack at Bütgenbach, the cooks left behind sacks of flour, large cans of whole coffee beans and crates of "D Rations" for the Schmetz family. But they took that special field kitchen with them. The ever-clever Marcel, however, had a plan.

Marcel knew the National Museum of Military History in Diekirch, Luxembourg had two identical, genuine field kitchens in their collection. He drove to the museum and asked to see the owner. An intrepid bargainer, Marcel relentlessly convinced the owner to trade one of the field kitchens for a Korean War mine detector. That field kitchen became an integral part of the first museum display, a reproduction of the Schmetz family kitchen. The replication—complete with life-size mannequins hand-built by Marcel—depicts the family and several D Company soldiers relaxing while awaiting their meal.

Replica of the Field Kitchen that was set up in the "Summer Kitchen."

The exacting details of the display represent the soul of the lifetime reverence Marcel had for the American soldier. In it, Henri, Sr. is presiding over the group. (The mannequin representing Henri, Sr. has a darker skin than the others. Marcel feels that this accurately portrays his father's dark, tanned complexion from spending much more time in the sun than the rest of the family). Two D Company cooks are preparing the meal with the field kitchen. (Which would have been outside the house in the "Summer Kitchen".) At the end of the table, Maria is sewing winter camouflage for the soldiers. Henri, Jr. is playing his accordion, and Marcel is engaged in an early version of the board game "Battleship" with one of the soldiers. Adorning the display is a metal soap dish left behind by the D Company commander's driver, Bill Lee. Lee's girlfriend Doris had sent him the almost indestructible container, with his name inscribed on the lid, to keep his soap clean in the field. Bill Lee would go on to marry that girlfriend after the war.

The Schmetz Kitchen, December 1944.

(Maria sewing, Soldier heating water, Marcel playing the Battleship game with a D Company Soldier, Henri, Jr. playing his accordion and Henri, Sr., excessively tanned from field work, per Marcel. The unnamed young women are neighbors who would frequently come to visit the American soldiers.)

Also prominently displayed was the popcorn cook pot "liberated" by Clayton Remaly on one of the Americans' "official business" trips to Aachen. Maria had made good use of that "new" pot over the following five decades.

Bill Lee's "Battlefield" soap dish.

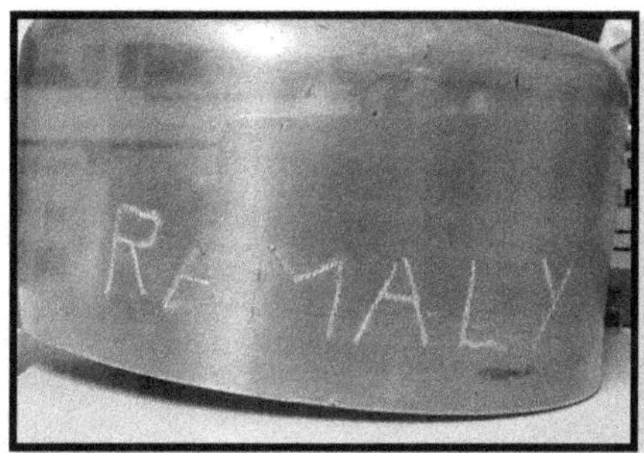

Clayton Remaly's "Liberated" cook pot.

Marcel and Mathilde were able to travel to America in 1996 for the Blue Spaders Reunion. Marcel could not remember the soldier who scratched his name on his mother's pot. He took a picture of the pot to the reunion and showed it around to the attendees. They immediately identified Clayton Ramaly, but he did not attend the reunion that year. Mathilde wrote a letter to Ramaly after returning home. They arranged to meet at the 1997 Blue Spaders reunion in Lancaster, PA.

Many of the previous reunions had been held in Bennie Zuskin's "neighborhood" in Virginia. But in 1997, it was held down the street from Sam Messina in Lancaster, PA. As planned, Clayton Ramaly met Marcel for the first time since December 16, 1944, almost 53 years earlier. They compared memories of those impactful two weeks or so. Both recalled Clayton helping with the chores every day, especially milking the cows. Clayton remembered that 11-year-old boy dressed in a reasonable facsimile of an American uniform saluting the men as they left for Bütgenbach. And Marcel explained the joy that pot had brought to him and his mother for so many years. He and Mathilde presented Ramaly with a set of ceramic cows named Aguesse, Marette and Finette, after the family cows he milked that cold December so long ago. Mathilde then invited him to come to visit Belgium and the museum, assuring him that, *"This time you can sleep in the house."*

The Red Ball Express

The Red Ball Express was a famed truck convoy system that supplied Allied forces as they rolled quickly through Europe from the D-Day beaches in 1944. To expedite precious cargo shipments to the front, tractor-trailer trucks emblazoned with red balls followed a similarly marked route that was closed to civilian traffic. Marcel had acquired one of the legendary Red Ball Express tractors in the 1980s, then acquired a genuine cargo trailer to accompany it in 1994. The tractor trailer became a centerpiece of the museum.

The Red Ball Express on the left.

Later, Marcel converted the trailer into a small theater where videos—made by Americans—are shown to tour groups. More than 700 veterans have autographed the truck, including two German veterans. The vehicle used to be driven to various commemorations, but it requires at least one liter of fuel per kilometer, a cost that has become prohibitive.

Veterans signatures on The Red Ball Express.

That first group of veterans returned home and quickly began spreading the word about a wonderful little museum in a small Belgium town that focused not only on displaying "war stuff," but also on telling the soldiers' personal stories. Their stories! Over the next few years, more and more WWII veterans learned of, and traveled to, the little museum in that small Belgian town. And with them, they brought personal, genuine artifacts as donations to the museum, artifacts that are priceless and irreplaceable. Each soldier was (and still is) treated like they were Eisenhower or Patton visiting. Possibly better, as they knew these were the men who did the bleeding.

A Nickname Is Born

Marcel and Mathilde were meeting so many wonderful people and making so many new American friends, it was only a matter of time before they were given a nickname. Of course, it would be bestowed upon them by an American veteran. Francis Chesnick, who fought with the 99th Infantry Division in WWII, had multiple reasons to visit the Province of Liege. In addition to returning to his Battle of the Bulge positions, his family had five men who had fought in the war. One, Anthony Chesnick, was killed in action and is buried at Henri-Chapelle Military Cemetery.

While corresponding with the Schmetzes, Francis noticed a repeating pattern: Mr. and Mrs., Marcel and Mathilde. He immediately thought of that famous American candy, and before long, a new name spread among the veterans. Thanks to Francis Chesnick, Marcel and Mathilde had become The M&Ms. Before long, they found themselves inundated with American advertising souvenirs. The innocent nickname is perfectly appropriate, as during WWII, 100% of Mars, Incorporated's M&M candy output was sold to the military to include in soldiers' rations and bolster morale.

The M&Ms in front of a small section of their namesakes.

The relationship with the Chesnick family opened a new means of continuing the M&M's mission. The 611th Graves Registration Company had buried more than 17,000 Americans in Margraten, Netherlands by the end of the war. As the unit prepared to return to the United States, the commander asked a local office worker if he could look after the graves. The man rallied the local population with an idea. Within a few months, every grave had been "adopted," and the locals adorned each grave with flowers. This program was soon formally organized across all American Battlefields And Monuments Commission cemeteries. Mathilde, in her fledgling English, asked Francis Chesnick if they could have the honor of adopting Anthony's grave. Of course, he agreed. It would be the first of 14 graves Mathilde and Marcel would adopt at Henri-Chapelle.

Preparing for Veteran's Day, November 2019.
(Mathilde and the author at one of her soldier's graves.)

The M&Ms were on a very limited budget, with no outside funding or grants, so advertising for the Remember Museum was mostly word-of-mouth. Initially, they charged only two euro to enter the museum, but only for Belgians and other international neighbors. American veterans and active-duty soldiers are never charged. Marcel says, "They paid more than enough in 1944." Almost 30 years later, the entrance fee is only 10 euro for adults and three euro for children. (As of July 2023).

Beginning with Ken Stanger in 1992, the M&Ms have hosted and fed visiting WWII veterans and their wives, sheltering them in their home and always refusing to take payment. They have hosted dozens and dozens of veterans, many of whom have visited several times.

Over the years, Bennie Zuskin, Sam Messina and Rocky Moretto became very close with Marcel and Mathilde. They visited often, bringing uniforms, photographs and other items to display in the museum. Bennie and Sam once spent three weeks with the M&Ms, helping Marcel install a modern bathroom on the second floor of the more than 300-year-old house.

Reuniting With An Old Comrade

Before he became a regular at the Remember Museum, Rocky had visited Belgium before. He traveled to Europe in 1984 with a few other Big Red One veterans for the 40th anniversary of D-Day and the Battle of the Bulge. After 40 years, he had hoped to reunite with "John the Belgique." Armed with only wartime photographs and the nom de guerre, Rocky enlisted the aid of local newspapers, radio and television to track down John.

Ultimately, they found him in a hospital in Liege, where he was recovering from a severe stroke. His real name was Jerome Bollen and his wife was at his side. He had been drifting in and out of consciousness. As Rocky entered the room, Jerome immediately became alert and grabbed Rocky's hand. Tears fell just as they had during their farewell in March 1945. *"When I got home,"* Rocky said, *"I received a telegram from the Division saying that Jerome had passed away. But before he passed, he had said my name—Rocky Moretto."*

Later, Rocky met with Jerome's wife, daughter and son, and was given Jerome's personalized bracelet as a memento of his comrade in arms. He treasured this as much as Marcel treasured his soldier's watch, gifted 40 years earlier.

"We Left That Tank In Belgium For The M&Ms"

In 1997, Bennie made one of his many visits. Marcel had long been aware of WWII armored vehicles that were being kept at Camp Elsenborn, a nearby Belgian military training post. Elsenborn was the same post the 26th Infantry Regiment had moved to after hurriedly departing Clermont on December 17, 1944. Over the years, the Belgian Army had used many of those old wrecks for target practice. Among the remaining historic vehicles was a tattered American M4 Sherman tank that had been weathering the elements for more than 50 years.

The camp's commander knew of the Remember Museum, and Marcel was scheduled to discuss with him the possibility of obtaining the tank for the museum's collection. Intrigued, Bennie asked if he could tag along. Upon meeting the Belgian camp commander, Bennie informed him that the American Army had *"left that tank in Belgium for the M&Ms."* The commander found it quite difficult to mount any sort of counter-argument, and Mathilde sent an official request through appropriate channels to the Belgian Army Headquarters in Brussels. In 15 days, the tank became the property of the museum. The challenge: how to get the elderly beast from Elsenborn to Clermont.

Two of Marcel's friends owned a transport company. They had visited the museum in the past. When they heard the story, they volunteered to move the tank to the museum. With that, just as he had done on so many automobiles over the years, Marcel began to work his restoration magic. It took him and a few friends more than 2,000 hours over nine months to painstakingly restore the exterior to its original condition. On September 11, 1998, the 54th anniversary of the liberation of Clermont, the Sherman was officially dedicated at the museum. It stands proudly, as it did so many years before.

Marcel's M4A4 Sherman before and after restoration.

In 2000, Belgium passed a law prohibiting private ownership of war weapons. Many WWII weapons were possessed by private citizens who, over the decades, had found them on their property or in the woods and forests immediately following battles. A great many of them had been passed down from generation to generation. Given that lineage, surrendering them to the government—to be destroyed—was a painful concept for many. The law also made it unlawful to sell these artifacts, which made donating them to a museum the next best alternative. It was through this legislative-inspired generosity that Marcel was able to expand the museum's weapons collection.

When the museum first opened in 1994, most of the WWII veterans were more than 70 years old. Life responsibilities, health and affordability made travelling back to Europe a luxury some could not afford. But word-of-mouth at WWII reunions in the United States, and good old-fashioned letter writing between friends, became a major catalyst for Remember Museum awareness, publicity and growth. Simultaneously, retirements, decades of savings and a sense they may not get another chance, inspired more and more veterans to return to Europe. The number of veterans making the pilgrimage to their old battlefields began to snowball. And many of them started bringing their own artifacts to the museum. Their precious personal artifacts and photographs, along with his own personal recollections, inspired Marcel to add more life-sized dioramas to the museum. Again, he would fabricate the mannequins—and even some of the war machines—himself. By 2000, Marcel had to add a third floor to the main building, just to accommodate the new acquisitions.

Bennie And Sam

Sam Messina and Bennie Zuskin were mostly inseparable during the war, but for Sam's two Purple Hearts, which had given him short reprieves from the front lines. He was stabbed in the shoulder and later was wounded in the elbow. Neither injury was severe enough to get home. Both men had made multiple trips to their former battlefields and to the museum. As the M&Ms got to really know these men, Marcel wanted to find a way to memorialize their good friends. So, he built a life-sized diorama of the two of them in one of their many fighting position iterations. Bennie and Sam purchased the Browning Automatic Rifle (B.A.R.) and donated it to the Museum to complete the display.

Diorama of Sam Messina and Bennie Zuskin holding the line near Bütgenbach. (The B.A.R. and disarmed hand grenades are real.)

On September 11, 1999, Bennie, Sam and Harry Wagner (another C Company veteran) attended Memorial Day ceremonies at the Henri-Chapelle Cemetery. They also spent time with Marcel and Mathilde before heading to Omaha Beach for the 55th Anniversary of D-Day. The Commanding General of the 1st Infantry Division at that time, Major General David Grange, provided transportation for their travel from Belgium to Normandy. General Grange had also become a friend of the M&Ms after visiting the museum several times.

On the Omaha Beach shingle where they had landed 50 years earlier.
(l. to r., Harry Wagner, Sam Messina, Jack Hathaway, Bennie Zuskin, unknown.)

The friendships Marcel and Mathilde developed over the years have paid great dividends. They have been invited to the United States dozens of times since they married. As of 2022, the M&Ms had made 25 trips to America, visiting many states. Many of those times, grateful WWII veterans and their veteran associations paid the airfare and lodging to have Mathilde and Marcel speak at their reunions. Those veterans also loved to show off their hometowns, and their grandkids, to their dear Belgian friends.

The Henri-Chapelle American Cemetery

Henri-Chapelle Cemetery was also a major draw for the men. It was the first time they would have the opportunity to "see" their friends since they died. By the end of the war, Vito Mastrangelo and his fellow Graves Registration soldiers had created 17,323 American, 191 Allied and more than 10,000 German graves at Henri-Chapelle. In 1947, the United States government appropriated more than $600,000 for a repatriation program of American remains. Next-of-kin were given four options for their loved one:

1. *Interred or reinterred in a permanent America military cemetery overseas.*
2. *Returned to the United States and internment in a National Cemetery.*
3. *Returned to the United States and internment in a private cemetery.*
4. *Reinterred in the country they are currently interred, or returned to a foreign country, the homeland of the deceased or the homeland of the next-of-kin.*

In the end, 56% of families requested that the remains of their loved ones be returned to the United States for burial in a National Cemetery or private cemetery. The remaining 7,992 rest in perpetuity at Henri-Chapelle.

Remembering Bennie Zuskin

No American veteran visited the Remember Museum and the M&Ms more than Bennie Zuskin. He had dedicated their museum on its opening day, helped them obtain the Sherman tank, dedicated the tank and brought numerous veterans into the fold. He did more than anyone to highlight and personify the Big Red One's selfless sacrifices for the Belgians, always shining the light on the men, never himself.

For decades after WWII, Bennie stayed in contact with the succession of Commanding Generals of the 1st Infantry Division who were stationed in Germany. So, for the 56th anniversary of the liberation of Belgium in 2000, the M&Ms arranged a special celebration, specifically with Bennie in mind.

Planning the event was almost as complicated as the Allied invasion of Normandy. The guest list included: the mayors of Aubel, Herve, Welkenraedt and Thimister-Clermont; the Commanding General of the 1st Infantry Division; the 1st Infantry Division Band; the Commandant of Camp Elsenborn and his predecessor; staff from the Supreme Headquarters Allied Powers Europe (SHAPE) and the Superintendent of the Henri-Chapelle American Cemetery. The guest of honor would be Bennie. And he and his wife, Maxine, ("Mac", Bennie's pet name for his wife), would be treated like royalty.

The M&Ms presided over the two-part event. Part one commenced at the 1st Infantry Division Monument in the town of Henri-Chapelle, newly restored by Marcel, Mathilde and their friends. Ceremonial wreaths were laid by Bennie and Major General John Abizaid, then Commanding General of the 1st Infantry Division. That was followed by a 21-gun salute, a sobering rendition of Taps and inspired words by Mathilde. Bennie Zuskin then gave an emotional speech.

A Man Of Few Words Speaks

"It is an honor for me to be here today as a 1st Division veteran of WWII in front of the newly restored monument by Mathilde and Marcel. At the beginning of World War II, the 1st Division had 14,851 men. From 1942 to 1945, there were 4,325 killed in action, 15,452 wounded in action and 1,241 missing in action. Throughout the war, there were 28,892 replacements, which made a total of approximately 50,000 men who served in the Division during World War II. Here at the Henri-Chapelle Cemetery, there are 626 soldiers from the 1st Division laid to rest; 254 of those soldiers were from the 26th Infantry Regiment, in which I served.

I was in Company C for the 26th Infantry Regiment on D-Day, June 6th, 1944. We had 240 men in the Company, which would never go over 200 again. By the end of the war there were over 850 men that served in Company C. There were only two men from Company C that went from D-Day June 6, 1944, to the end of the war May 8, 1945, without being wounded. I was lucky to be one of them. There are four more monuments like this one in Europe (with names of 1st Division soldiers). I remember only a few names:

Sgt Robert Allen —
Killed by a mortar in Diepenlinchen.

James Portar —
I was in the same hole when he
had both legs blown off.

Harry Bonar.

Edwin Doucet.

Ernest Delaney —
He was a medic. The medic had the worst job in the Army because when the shelling was going on, you were in your hole, but they had to get out and go to help the wounded.

There are names on this monument of men that I do not remember because they were replacements, they came in one day and were sometimes gone the next. I can remember when we got replacements from the Air Forces. When they were told to dig in with someone by the Sergeant, they looked like they did not know what he meant. But it did not take them long to find out what dig-in meant. The living are not the heroes of today, they are the ones who are laid to rest in the Henri-Chapelle Cemetery. I was not scared of the small arms fire, for most of the time you could see where it was coming from. It was the artillery and mortars that I could not see. That you feared the most. The worst part of the Hürtgen Forest were the tree bursts.

I would like to thank all of you for being here. It shows your respect for this monument and the men whose names appear here. God bless the Belgian people."

Part two was held at the museum, which opened its doors to all attendees. Here, Bennie was christened "Citizen of Honor of the Commune of Thimister-Clermont," the first such honor bestowed in the 800-year history of Thimister-Clermont. In her speech, Mathilde declared, *"It is through Bennie that we honor the bravery of all American soldiers who died or are still living."* She then proclaimed Bennie the "Honorary President" of the Remember Museum 39-45.

Bennie and Maxine ("Mac") Zuskin at a Blue Spaders reunion.

On October 12, 2005, Mathilde made one of her many phone calls to the Zuskin home. For some reason, Mathilde had felt a special urge to call that day. She knew that the asbestosis Bennie had from the years of labor in the Virginia shipyards had turned cancerous. Mac answered the phone and they quickly caught up on happenings. Bennie was sleeping, but Mac insisted on waking him, as he would never want to miss a call from his "younger sister." Mathilde gave him a pep talk and implored him to exercise and take care of himself. *"We want to visit you next year,"* she wished out loud. Then she added hopefully, *"And you have to visit us one more time."*

"Yes, Darling," he said, sounding uncharacteristically serious as he spoke. It was the first time he had ever uttered a term of endearment around the M&Ms. Bennie Zuskin passed away the next day at the age of 80. ✪

10

ENGINEERS, AIR CORPS & OTHERS

IN JUNE 2002, MATHILDE ANSWERED A KNOCK at their door. As with Ken Stanger 10 years earlier, it was an American she did not recognize. This one was too young to be a WWII veteran. It was John Tait, Jr., from New Jersey, who was retracing his father's footsteps through Europe with the 1st Infantry Division in 1944-45. His father had been in a jeep that stopped in Clermont on September 11, 1944.

Staff Sergeant John Tait was advancing with his unit, the 1st Engineer Combat Battalion of the 1st Infantry Division along with PFC Bill Clements and four other soldiers. John Jr. produced a copy of a photograph, taken so long ago, of a jeep surrounded by civilians. Mathilde's jaw dropped. It was the photograph taken by Joseph Baguette on Liberation Day 58 years earlier!

For all those years, many families in Clermont had copies of the famous photograph. They could name all the townspeople, but they could never identify the names of their liberators. In 1998, Bill Clements attended the annual 1st Engineer Combat Battalion reunion and met John Tait, Jr. The younger Tait's father had passed away a few years earlier. Clements gave the younger Tait copies of all the pictures he had from the war, and the two of them worked out the other four soldiers' names. The townspeople would finally know.

Armed with this new trove of information, Mathilde placed a phone call to Bill Clements in Saratoga Springs, NY. Before long, plans were being made to have Bill, his wife and son travel to Clermont to celebrate the Liberation of Clermont on its 59th anniversary. Bill became "King for a Day" and was treated like royalty. A genuine WWII American jeep drove Bill down the same Rue René Rutten route that those "famous" engineers had driven 58 years before. They stopped at the same house and recreated the original photo as closely as possible. The surviving civilians took their original positions. Unfortunately, like so many of the M&Ms' veteran friends, the five Combat Engineers who had accompanied Bill Clements in 1944 had passed and could not celebrate with them.

Original image (left) of the Liberation photo with Bill Clements' Jeep. The anniversary photo (right) when Clements returned in 2002.

The 1st Engineer Combat Battalion still has reunions, but newer generations of soldiers fill the ranks. The organization made the M&Ms lifelong members, to thank them for their work to keep the WWII Engineers stories and memories alive.

The exploding volume of correspondence between Marcel, Mathilde and their heroes from the Big Red One and other units ultimately inspired a few of the veterans to arrange a special visit to the museum for its 10th anniversary in 2004. Three old soldiers and their families visited the M&Ms for a joyous, tearful reunion. Naturally, Bennie Zuskin led the charge, accompanied by Rocky Moretto and Frank Vargo. Though only a Private First Class during the war, Bennie was the General in this group. In her usual comprehensive way, Mathilde arranged a multi-pronged celebration. One part was held at the Henri Chapelle Cemetery. There Mr. David Atkinson, the Superintendent at the Cemetery at the time, donated to the museum the huge American flag that had flown over the fallen soldiers.

On September 16, 2004, the week of the great celebration to

commemorate the 60th Anniversary of Belgium's Liberation, Henri Schmetz, Jr. passed away. Those three American veterans attended the celebration—and attended Henri's funeral.

Picklepuss

Marcel still has dreams about some of the events that marked his childhood. One in particular was of the B-17 bomber that he watched explode and crash in 1943. He always wondered what had happened to that crew.

Marcel researched many of the US military records that had become available to the public over the decades. He learned that the plane was enroute to bomb the Messerschmidt factory in Regensburg, Germany. It was in the "coffin spot," the last plane in the lowest echelon in a large bomber formation. This made it more vulnerable to enemy fighters.

History would tell that the aircraft was nicknamed Picklepuss, an endearment the pilot, Robert Knox, called his wife, Dorothy. Picklepuss was attacked and had one of its four engines damaged. Limping, it could not keep pace with the rest of the formation and soon was alone over enemy territory. Knox turned the plane around in an attempt to make a run for friendly territory, but it was intercepted by two more fighters. The aircraft was mortally wounded, and the crew attempted to bail out. Only four made it out. All were either injured while in the aircraft or when they hit the ground. All four survivors became prisoners-of-war. The other six crew members perished, including Robert Knox.

The dead were temporarily buried near the crash site, but were later disinterred and reburied, most likely by Staff Sergeant Vito Mastrangelo's Graves Registration Company at the Henri-Chapelle Military Cemetery. In 1947, five of the dead were repatriated to the United States at the request of their families. One remains at rest at Henri-Chapelle, the pilot Robert Knox. His grave was adopted by Marcel and Mathilde. In 2006, Marcel sought and obtained help from the town of Plombières to build a memorial to the crew at the exact spot where he and his father saw those flames. It was dedicated that same year.

Picklepuss and her crew.

U.S. Air Force personnel from Ramstein Air Force Base at the memorial dedication in Plombieres.

Keeping Memories Alive For New Generations

It cannot be overstated how much Marcel and Mathilde want others to understand and appreciate the sacrifices American soldiers had made to liberate Belgium and restore its freedom. Over the years, they have invited busloads of local school children to tour the museum and learn their region's history, directly from instructors who could bring it to life better than any contemporary history teacher. Naturally, the weapons and accoutrements of war are alluring to the young. But even the "busiest" children quickly give rapt attention to the amazing stories of mere boys doing heroic things, as narrated by Marcel and Mathilde.

The children quickly realize that—in another time—those boys could easily have been their older brothers. They also come to understand that the experiences Marcel shares are exactly what their grandparents or great-grandparents likely experienced. Thousands of school children have been enlightened over the decades, opening doors to many new multi-generational dialogues. Some of these children have also been lucky enough to meet the American heroes who liberated their grandparents. Mathilde always tries to arrange for a group of students to attend the museum when she knows her "big brothers" will be in town. As a special bonus, Mathilde always gives the children their own pack of M&Ms, which have been provided by their American friends.

9/11

Early in the new millennium, a different mission arose for the M&Ms. The 9/11/2001 terrorist attack on America cast a pall on the Liberation Day celebrations of Thimister-Clermont and the surrounding communities. To them, it felt inappropriate to celebrate that day, knowing their American friends had experienced another day that "would live in infamy." The 9/11 attack launched two wars—Iraq and Afghanistan—that would go on to become the longest wars in American history.

Many of the casualties from those wars would be transferred to Landstuhl, Germany, home of the only American military hospital in Europe. Many of the injured, if they could ultimately return to duty, would stay in Germany as they recovered from their wounds. Landstuhl is roughly a two-hour drive (240 kilometers/149.1 mi.) from the museum, a short distance that would spawn a new generation of American warriors who would enter the hearts of the Mathilde and Marcel.

Van loads of wounded soldiers, sailors, airmen and Marines, along with their medical caregivers, visited the museum. Each of them was given the VIP treatment. Following a guided museum tour, each group ultimately ends up in the M&Ms' living room, now known as the "Gathering Room". There the men and women are often toasted with fresh-made waffles and a 12-ounce Jupiler (Marcel's favorite pilsener), as they sit on the "Bench of Honor." It is the same bench that had been "blessed" by the D Company soldiers in Henri and Maria Schmetz's kitchen so many years ago. These new heroes would come to be looked upon as the M&Ms' "children and grandchildren," not their "big brothers" of another era.

Many of these medics would return to help Mathilde clean the museum, or help Marcel refurbish the Sherman tank standing guard outside. That Sherman had originally arrived without the .50-caliber machine gun that should have been mounted on the top of the turret. So, Marcel fabricated—from scratch—a "machine gun" to attach to the gun mount. In yet another testament to Marcel's artistic prowess, someone tried to steal the gun, thinking it was real. Naturally, Marcel had anticipated the "compliment," and had permanently welded the realistic reproduction to the tank.

Paying It Forward

The United States Air National Guard contains all the military's air-refueling capability, outside of the active Air Force. To support Operation Iraqi Freedom and Operation Enduring Freedom, the USANG began rotating air-refueling units to a NATO base in Geilenkirchen, Germany. The base is less than an hour drive from Clermont, and the Remember Museum. A liaison officer became aware of the M&Ms' mission, and soon a trip to the museum unofficially became part of the airmen's rotation. The Guard units are a part of an individual U.S. state's National Guard, and many of the visitors often bring unique mementos of their state to the museum. For many of those young military members, it was their first opportunity to learn of the real sacrifices of the Greatest Generation. And from Belgians no less!

The 171st Air Refueling Wing from Pittsburgh, PA did their part to "pay forward" the museum's mission. A bible left behind in 1944 in the town of Spa by an American soldier, likely in a rush to move out with his unit for the front lines, was found and kept by a Belgian family. Over time it was passed through the family to Serge Fafchamps, a local policeman. For years, the man wanted to find a way to get it back to the soldier or his family. Like so many Belgians he knew of the sacrifices the American soldiers had made and wanted to let it be known that they were not forgotten so many decades later. He also hoped it would bring some small closure and happiness for the family. "Private Millard Weekley" was inscribed on the inside of the bible and Fafchamps had made a diligent search for the owner and his family, but had reached a dead end. Or so he thought. He heard of the Remember Museum and how Marcel and Mathilde had communicated with so many veterans and their families. Soon the M&Ms

were using their extensive American network to find Private Weekley or his relatives. It was not long before someone found his daughter, Paula Ferrell, in Coraopolis, PA.

It happened that Major Jim Moretti and his brother Sergeant Dan Moretti were crew members of a KC-135 tanker plane that had previously rotated to Geilenkirchen, and was returning for another mission. Like the Moretti brothers, the American soldier who had left the bible behind was from Pittsburgh...and their air base was right next to Coraopolis! The brothers agreed to help find the family and return the Holy Book.

They tracked down Paula Ferrell, and invited her to a special ceremony at the 171st Air Refueling Wing. There the Bible was returned to the family. It now sits on a nightstand next to her bed, opened to her father's handwriting. While Millard Weekley never spoke much of his war experiences, he now speaks to his daughter nightly.

The M&Ms with the Moretti Brothers at the Air Base in Geilenkirchen l. to r. Major Jim Moretti, the M&Ms, Captain Dan Moretti.

More Than A Museum

The Remember Museum is the physical manifestation of the M&Ms' reverence for America and her veterans. But their love extends beyond the obvious WWII connections. They have also opened their hearts and their home to today's wounded warriors and their medical teams. They have been powerful ambassadors between Belgium and the United States, especially the military. This has not gone unnoticed. General Bantz J. Craddock was the Supreme Allied Commander, Europe (SACEUR) from 2006–09. The four-star general recommended the Schmetzes for an American medal rarely given to civilians, let alone foreigners. In 2007, the Schmetzes were awarded the Military Outstanding Volunteer Service Medal.

REMEMBERING MORE THAN SOLDIERS

As you drive up to the Remember Museum, you cannot miss the 32-ton M4 Sherman tank and the reinforced-concrete "dragon's teeth" in front of it. By the front door there stands a steel and concrete "one-man pillbox." Memorial plaques hang on the outside wall dedicated to the Big Red One and the 607th Graves Registration Company.

Monument to the war animals.

There also stands an unusual war monument—developed by Marcel and Mathilde—dedicated to those who died in "forced service." Growing up, Marcel was surrounded by animals on his parents' farm. They were always treated well. Marcel was appalled at the slaughter of animals during the war. Some were farm animals caught in crossfires or shelling. But many were pressed into service, blindly obeying their masters with no understanding of war. The Germans used millions of horses throughout the war. Dogs and pigeons were also used in significant numbers. Many suffered greatly, rarely receiving medical care. Marcel always wanted to honor their sacrifice. On September 11, 2009, the monument was dedicated. It includes a horse, a dog and a pigeon.

Marcel has taken another step closer to memorializing "in perpetuity" the veterans who have had the most impact on the Museum's mission. On one of Bennie Zuskin's many visits, he was sitting out front of the Museum with Marcel. The Museum had been open for a couple years and Bennie was thinking of how he could help his dear friend keep their important mission alive. At once he announced he would make a cash donation to the museum, in addition to the personal items he had already given. In turn, Marcel would inscribe his name and unit on an exterior brick. He would then encourage his fellow veterans to do the same. It was not long before Marcel ran out of bricks on the front wall and moved around to the side of the building as donations poured into the Museum. Some of the aging veterans could not afford cash donations but were happy to send irreplaceable items for display. The sad reality is that it is likely there will not be many more engravings. ✪

Marcel pointing to one of the memorial bricks. (Charlie Amero was a D Company hero who stayed on the Schmetz farm in December 1944.)

EPILOGUE

MARCEL AND MATHILDE SCHMETZ HAVE SPENT countless hours building, expanding and maintaining their beloved museum. They have spent even more hours hosting, honoring, remembering, and communicating with their heroes and their families.

Each year, Mathilde sends nearly 700 Christmas cards to her American friends. Each includes a full-color letter highlighting events involving their heroes from the past year. The M&Ms receive at least that many cards in return. Over the years, they have collected boxes of correspondence. Many of the letters were handwritten by WWII veterans in their 80s or 90s. Others came from nonagenarian wives — or children themselves in their 60s and 70s—informing that their soldier had passed on. Mathilde has every bit of it organized by unit and soldier.

The museum keeps Marcel and Mathilde very busy. But their mission goes beyond the building, its artifacts and the soldiers' stories. They have formally adopted 14 graves at the Henri-Chapelle American Military Cemetery. They adorn the fallen soldiers' graves with flowers and flags on pertinent holidays such as Memorial Day, Veteran's Day and each soldier's birthday. Photos are taken and sent to the families. The life of the soldier is studied and learned, just like the many veterans who have become the fabric of the museum. The M&Ms purchase 1,000 roses each year for Memorial Day. They place one on each grave at Henri-Chapelle, rotating cemetery sections. It takes seven years to ensure every soldier receives their own rose for Memorial Day.

Marcel and Mathilde have a special relationship with the American Battle Monuments Commission (ABMC). The ABMC was established by the U.S. Congress in 1923 and is an agency of the executive branch of the federal government. As the official guardian of America's overseas commemorative cemeteries and memorials, it honors the service, achievements and sacrifice of U.S. Armed Forces.

In addition to Henri Chapelle, the M&Ms also work with the Ardennes American Cemetery just southwest of Liège. They direct schoolchildren to the west of Thimister-Clermont to the closest cemetery to learn of the Americans' sacrifice. In return, the Cemetery staff notify the M&Ms of any next-of-kin who plan to visit their soldier. They are invited to visit the museum, sit in the Gathering Room, talk and ask questions about their loved one. This is especially true for the WWII orphans who never had the chance to meet their father. The M&Ms are well known to the American War Orphan Network (AWON) and have provided closure for many orphans of all ages.

Vito Mastrangelo

I had the honor of interviewing Tech Sergeant Vito Mastrangelo, who had landed at Omaha Beach on D-Day with the 607th Quartermaster Graves Registration Company. Despite severe hearing loss and chronic pain, Vito spoke with me for more than an hour.

When our call ended, Vito agreed to a follow-up call to provide even more insights. When I called a few weeks later, his caretaker answered. She sadly informed me that Vito had passed away the week before. He was just shy of his 99th birthday.

Never Forget.

To this day, Marcel, offers his American-veteran guests fresh croissants and chocolate for breakfast when they come to stay, just as Maria fed her sons a piece of D ration chocolate bar and fresh baked croissants made from the GI flour left behind in December 1944.

Marcel also has some of the original coffee beans left behind as a gift to his parents so many years ago. On one of the author's visits, Mathilde opened the precious 80-year-old coffee can and the amazing aroma of "fresh" beans was still noticeable as soon as that lid cracked open.

Mathilde with the 80-year-old coffee beans.

These days Marcel looks forward to the visits from Active Duty, Reserve and National Guard soldiers, airmen, sailors and Marines. When they walk through the door, his mind reverts right back to that cold day in November, 1944 when the soldiers of D Company, 1st Battalion, 26th Infantry Regiment, 1st Infantry Division trudged through the door of his parent's home. He is sometimes surprised at how little these young Americans know about their predecessors and how much they learn on their visits to the museum. And he looks forward to educating many more. *"No matter what side of the ocean we are on,"* he says, *"or how old we are, we must remember."*

A Doctor After All

The M&Ms know their soldiers' stories so intimately, I often feel that I interviewed the men personally, though they had passed away years ago. At the Blue Spaders reunions organized by Bennie and Maxine Zuskin, several of which the M&Ms attended, soldiers could speak freely of the unimaginable horrors they had experienced. The Schmetz home became a similar "safe haven" for the men to speak candidly with their dear friends and veterans.

On one of their last visits Bennie Zuskin and a few fellow C Company veterans found themselves sitting and reminiscing in the M&Ms' home, on the Bench of Honor, of course. The veterans knew this could be their last visit. In a moment of quiet reflection, Mathilde rued the fact that she had never gotten the chance to become the physician she had dreamed of being as a young woman. After a thoughtful pause, Bennie, the man of few words, leaned forward and said, *"Mathilde, you and Marcel are the best doctors us veterans ever had."*

The M&Ms on The Bench of Honor, November 2021.

WHAT THE FUTURE HOLDS

Today there are few WWII veterans of the European Theater left, and even fewer who are healthy enough to make the long trip back to rural Belgium. In 2023, even the youngest will be in their late 90s. Marcel Schmetz, himself, turned 90 in June 2023. Mathilde went all out for the celebration, with roughly 200 people in attendance. An American flair was only to be expected. Hot dogs, hamburgers, baked beans and American beer were enjoyed by all. Twenty American friends helped to prepare and serve the food and drinks. The main attraction was an Elvis impersonator who sang from the Red Ball Express "stage" with the trailer's driver side lowered. The M&Ms retire early and rise early but that night was an exception.

By his own admission, Marcel is slowing down. However he still does most of the maintenance and repairs at the museum and leads tours in French. He will not be able to keep this pace up for long.

There is no solid legacy plan for the Remember Museum 39-45. No one in the Schmetz family can assume control of the operations or tell the stories. The M&Ms' greatest fear is that one day their museum will cease to exist and their soldiers will be forgotten.

To date, the M&Ms have approached several organizations they believe could carry on their mission without compromising their vision. The McCormick Foundation, headquartered in Chicago, is one of them. Robert R. McCormick was an officer in the Big Red One during WWI and his foundation is responsible for the 1st Infantry Division Museum in Wheaton, Illinois. The idea of a sister museum on the very battlefields where the Big Red One fought, died and are still considered liberators and heroes by all generations puts the McCormick Foundation in high consideration.

Another is the King Baudouin Foundation, headquartered in Brussels, with an office in New York City. Their activities "aim to foster sustainable and positive change in society, in Belgium, in Europe and around the world."

The M&Ms have dedicated their lives to remembering a generation of brave young men. As time passes, they hope others will join them in keeping those memories alive.

CHARLIE MIKE: ***Continue the Mission!***

✪

THE END.

ACKNOWLEDGMENTS

So many have contributed to this labor of love. I was extremely fortunate to find several local Belgians who were alive during the war and old enough at the time to understand what they and their families were enduring.

Charles and Maria Putters were so kind to allow me into their home (Marcel Schmetz's childhood home) on multiple occasions to interview and take photographs. Collette (Putters) Rogister provided details of her family's experiences with the 607th Graves Registration Company on their farm. Georges Aalberts, through his mother's and grandmother's first-hand accounts and photographs, detailed the horrible events that occurred in Andrimont the day before their liberation. Alice Didden provided a fascinating account of her and her family's annexation and violent liberation. Stephen Fry, an American living in Belgium after meeting his Belgian wife on a Peace Corps mission, drove me all over Liege Province and provided nuanced translation for many interviews. Those included Alice Didden, Lucie (Dedoyard) Braggard, Renee Detry and Barbara Simons.

Ivo Van Meerbeek provided interpretation for interviews with Georges Aalberts, Roger Schmidz and Paul Hardy. These latter two interviews, though not related to this story, gave me a better understanding of the occupation and annexation from 1940-1944. There was Emma Muller, a vivacious 99-year-old whose own captivating story (figuratively and literally, she was a prisoner of the GESTAPO for five months) could be a book on its own. Mathilde Schmetz provided translation for that lively interview. Yves Bastin, a journalist for the Verviers Newspaper, Sudpress, provided the original manuscript of *The Official Account of What Happened, 1939-1940, a definitive telling of how and why Belgium entered WWII.* Christian de Marcken (age 95 at the time of this writing), whose father was an

American and was the General Manager of Hammond Organ operations in Europe. He married a Belgian and they raised Christian and his siblings in occupied Belgium. He is a good friend of Marcel and Mathilde and his ongoing encouragement to tell their story is much appreciated. Albert Trostorf, the Mayor of Merode, Germany, and published expert on the Hürtgen Forest campaign, was so kind to review the manuscript. Special thanks to William C.C. Cavanaugh, a published expert on the Battle of the Bulge, for introducing me to Marcel and Mathilde for the first time. Mireille Piquereau, who was 18 years old in 1944, visited the American General Hospital in Verviers and would hold vigils at the bedside of dying soldiers. She held their hands so that they would not die alone.

Of the soldiers, only Vito Mastrangelo was alive to be interviewed at age 98. The families of the other soldiers were very gracious in sharing memories and photographs. They include Philip Clements, Karen (Zuskin) Norris, Maxine Zuskin, Jill Moretto and Louis Muhlberg. Special thanks to Rick Amero, who shared the personal research he undertook to learn of his father's journey from enlistment to being severely wounded near Remagen, Germany ending his war in March 1945.

Michael Karpovage of Karpovage Creative designed the exquisite maps and charts that aid in bringing the reader into the world of Marcel Schmetz and helping to visualize and clarify the historic movements of the military units involved in the story. See more of his work at *KarpovageCreative.com*.

Rick Radermacher, a principal of Crabb Radermacher, and a good friend since first grade, provided editing pro bono. However, after visiting Remember 39-45 and being with the M&Ms he was so smitten by their mission I thought he was going to pay me! Thank you, old friend! ✪

APPENDIX A

THE SOLDIERS

AMERO, CHARLES (CHARLIE)
Born: April 15, 1924, Essex, MA
Drafted: May 10, 1943
Unit: D Company, 26th Infantry, heavy machine gunner
Wounded: March 1945, near Remagen; artillery fragments to his back and leg with nerve injury
Married/Children: Kathryn, Daughter Barbie, Son Rick, Daughter Kathy, Son Scot, Son Robbie
Occupation: Foreman, Battery Manufacturing Company
Died: July 11, 2003

AVON, CHARLES (CHUCK)
Born: July 13, 1919, New Philadelphia, OH
Enlisted: December 7, 1942
Unit: D Company, 26th Infantry
Wounded: March 1945, leg/foot wound
Awards: Purple Heart presented by Actress Tallulah Bankhead
Married/Children: Alice, Son William
Occupation: Inspector for metal rolling/stamping company
Died: October 6, 1988

BLETT, ROBERT (BOB)
Born: September 16, 1923, Worcester, MA
Drafted: February 1943
Unit: HHC, 1st Battalion, 26th Infantry, Assigned to C Company, combat medic
Wounded: November, 1944 at Merode; shrapnel wounds in the shoulder and back. (Friend and

fellow medic, Jimmy Carr, and his Platoon Leader Lieutenant Brooks (first name unknown), plus one other were KIA by same blast)
Married/Children: Claire, Daughter Sharon
Occupation: US Air Force radar operator
Died: July 30, 2018
Awards: Bronze Star for killing a tank; Purple Heart for wounds sustained near Merode
Other: Transferred to the US Air Force after WWII and served more than 20 years

CARR, JAMES (JIMMY)

My deepest sympathies to the Carr family, I was unable to find a photograph of Jimmy.
Born: April 7, 1923
Drafted: February 9, 1943
Unit: C Company, 26th Infantry, medic
Died: November 24, 1944, KIA

CLEMENTS, WILLIAM (BILL)

Born: September 27, 1923, Saratoga Springs, NY
Drafted: December 1942
Unit: Company A, 1st Engineer Combat Battalion
Wounded: Twice, Battle of the Bulge and crossing the Rhine River
Married/Children: Flora, 3 children
Occupation: Carpenter, cabinet maker
Died: November 16, 2012, Saratoga Springs, NY
Awards: 5 Bronze Stars, 2 Purple Hearts
Other: Was in the jeep that liberated Clermont on September 11, 1944

LEBDA, JOHN

Born: October 29, 1919, Miller Run, PA
Drafted: January 19, 1942
Unit: D Company, 26th Infantry
Married/Children: Caroline, 1 Daughter, 2 Sons
Occupation: Owned/operated a service station
Died: February 21, 2012, Baden, PA

LEE, WILLIAM (BILL) G.

Born: December 27, 1923, Centralia, IL
Drafted: November 25, 1941
Unit: D Company, 26th Infantry
Married/Children: Doris, 4 children
Occupation: US Postal Service postmaster
Awards: Bronze Star with two Oak Leaf Clusters

Mastrangelo, Vito

Born: May 2, 1924, Visalia, CA
Drafted: February 1943
Unit: 607th Graves Registration Company, processed and buried war casualties
Married/Children: 1946 Anne, Daughter Jean, Son John; 1979 Theresa
Occupation: Cabinet maker
Died: December 9, 2022

Messina, Sam

Born: December 13, 1924, Lancaster, PA
Drafted: February 1943
Unit: C Company, 26th Infantry, BAR gunner
Wounded: Twice, stabbed and shot
Married/Children: Anna, Son Louis Muhlberg, Daughter Patricia
Occupation: Electrician
Died: January 26, 2013
Awards: Purple Heart with Oak Leaf Cluster

Moretto, Rocco (Rocky)

Born: June 20, 1924 New York City, NY
Drafted: February 12, 1943
Unit: C Company, 26th Infantry, Squad Leader
Married: Monica, Son John
Occupation: Manager for Amtrak
Died: August 26, 2018
Award: Bronze Star

Ramaly, Clayton

Born: December 8, 1923 Palmerton, PA
Drafted: February 4, 1943
Unit: D Company, 26th Infantry, 81mm mortarman
Occupation: Machinist
Died: November 23, 1999
Awards: Bronze Star

Stanger, Kenneth

Born: December 25, 1922, Ogden, UT
Unit: Cannon Company, 395th Regiment, 99th Infantry Division
Married/Children: Evona
Occupation: Accountant; owned shooter store
Died: June 24, 2011
Awards: Bronze Star

Wright, Robert (Bob)

Born: N/A
Unit: C Company, 26th Infantry; Squad leader, rifleman
Died: KIA, January 24, 1945

Zuskin, Bennie

Born: June 25, 1925, Newport News, VA
Drafted: October 1, 1943
Unit: C Company, 26th Infantry, assistant BAR gunner
Married/Children: Maxine, two children
Occupation: Shipyard worker
Died: October 15, 2005
Awards: Two Bronze Stars

1-26th Infantry Regiment, Company C Reunion – May 1987.

APPENDIX B

M&M'S VETERAN CORRESPONDENCE LIST

Adams	Herb	82nd AB 504th P.I.R.	D Co
Aldrich	Bill	36th Fighter Gr.	
Aldrich	Frank	8th AF 305th Bomb Gr.	
Alford	Barney	106th Inf. Div. 589th FA	A Battery
Amero	Charles	1st Inf. Div. 26th Regiment	D Co
Angert	James	99th Inf. Div. 393rd Regiment	
Antonio	Ben	75th Inf. Div. 375th Medic.	A Co
Asselin	Edward	197th AA AW Bn	
Augustine	Gabriel	78th Inf. Div. 303rd Med. Bn	C Co
Bailey	David	106th Inf. Div. 422nd Regiment	F Co
Bailey	Cletis	84th Inf. Div. 333rd Regiment	K Co
Baker	Roy	1st Inf. Div. 26th Regiment	D Co
Barre	Dorothy	16th General Hospital	
Barts	Wilmer	924th FA Bn	
Baszner	Jerry	130th General Hospital	
Beach	James	82nd Airb. Div. 504th P.I.R.	HQ Co
Beeman	Mark	99th Inf. Div.	
Benefiel	Philip	99th Inf. Div. 393rd Regiment	A Co
Berry	Ben	863 QM Fum. & Bath Co	
Bicknell	Henry	5th Arm. Div. 10th Tank Bn	Serv Co
Biehl	Dick	1st Inf. Div. 26th Regiment	B Co
Black	William	75th Inf. Div. 289 Regiment	G Co
Bois	Paul	78th Inf. Div. 310th Reg	B Co
Bonacker	Marlyn	384th Bomb Gr.	
Bonner	Philip	159th Eng. Cmbt Bn	A Co
Boone	Stewart	99th Inf. Div. 924th FA	
Bourland	Clayton	740th Tank Bn	Service Co
Bowers	Howard	99th Inf. Div. 394th Regiment	D Co

Bowman	Rich	704th Tank Destroyer Bn	B Co
Bradicich	Bob	28th Inf. Div.	
Bradley	Harold	740th Tank Bn	aC Co
Branch	James	2nd Inf. Div. 38th Regiment	HQ Co
Brandow	Stuart	78th Inf. Div. 310th Regiment	D Co
Brockette	Marlin	1st Inf. Div. 26th Regiment	C Co
Bronson	Glenn	99th Inf. Div. 393rd Regiment	1st Bn
Brookie	J.B.	740th Tank Bn	HQ Co
Brooks	Stewart	555 SAW Bn (Radar)	
Brunger	Bill	78th Inf. Div. 289th Inf. Div.	B Co
Bucci	Alfred	99th Inf. Div. HQ	Art. Battery
Bull	Ivan	99th Inf. Div. 395th Regiment	L Co
Bull	Ivan	99th Inf. Div. 395th Regiment	L Co
Busbin	James	16th Reinforc. Depot	
Butler	Harry	106th Inf. Div. 424th Regiment	
Carter	Darrell	99th Inf. Div. 393rd Regiment	Med. Det.
Chaitt	Art	1st Inf. Div. 16th Regiment	HQ 3rd Bn
Chauvin	Lloyd	78th Inf. Div. 311th Regiment	M Co
Chesnick	Francis	99th Inf. Div. 393rd Regiment	A Co
Chipp	Harry	3rd Arm. Div.	
Christensen	Clayton	99th Inf. Div. 324th Eng.	A Co
Ciampa	George	607th Graves Regiment Co	
Cicchinelli	Joe	551st Prcht Bn Scout	A Co
Clark	Dick	740th Tank Bn	
Clements	Bill	1st Inf. Div.	1st Eng Cmbt Bn
Coffman	Warren	1st Inf. Div. 26th Regiment	C Co
Colbert	Hugh	106th Inf. Div. 422nd Regiment	B Co
Colby	Stan	99th Inf. Div. 393rd Regiment	K Co
Cole	Bob	740th Tank Bn	HQ Co
Cole	Galen	5th Arm.Div. 46th Inf. Bn	B Co
Cooley	James	106th Inf. Div. 423rd Regiment	D Co
Copeland	Linwood	643rd Tank Destroyer	A Co
Costello	Bill	1st Inf. Div. 26th Regiment	C Co
Coulter	Jesse	99th Inf. Div. 394th Regiment	D Co
Cross	Charles	Tuskegee Airman	
Crum	Dick	1st Inf. Div. 26th Regiment	C Co
Currey	Francis	30rd Inf. Div.	
Cutburth	Lee	174th FA Bn	
Dalton	Jack	86th Chemical Mortar Bn	C Co
Dauner	Jack	9th I.D. 60th Reg + AF	K Co

Davis	Dwight	740th Tank Bn	HQ Co
Davis	Dorothy	Army Nurse	
Dayton	Edie	740th Tank Bn	HQ Co
De Santis	Joseph	75th Inf. Div. 291st Regiment	
Dewing	Rose	Army Nurse	
Dilbeck	Larkin	740th Tank Bn	C Co
Diller	Robert	1st Inf. Div. 26th Regiment	634 TD Bn
Dixon	Sam	1st Inf. Div. 26th Regiment	G Co
Donald	Dick	Navy	
Dreher	Bill	78th Inf. Div.	F Co
Driggs	Walt	8th Inf. Div.126th Regiment	I Co
Drobnich	Bill	558th AAA (AW)Bn	Battery D
Dudasko	George	773rd FA Bn	
Duff	Hector	7th Armored Division	
Dusek	Melvin	99th Inf. Div. 394th Regiment	K Co
Dusseau	Paul	99th Inf. Div. 393rd Regiment	
Echmalian	Edward	557th Ord. Hvy Mnt Co	Tank Co
Edds	Harvey	28th Inf. Div. 110th Regiment	Battery B
Eldridge	Robert	106th Inf. Div. 422nd Regiment	G Co
Ellis	Paul	75th Inf. Div. 290th Regiment	K Co
Eriksen	René	9th AF.	
Erskine	Bob	78th Inf. Div. 303rd Med. Bn	D Co
Farmer	Marvin	29th Inf. Div. 116th Regiment	
Fary	Ray	82nd ABN Div.	80th ABN AA Bn
Fenniman	Arthur	558th AAA (AW)Bn	Battery D
Field	Dick	551st PIB	B Co
Finfroch	George	1st Inf. Div. 26th Regiment	C Co
Fleckenstein	Bob	740th Tank Bn	A Co
Foehringer	Roger	99th Inf. Div. 924th FA Bn	Service Battery
Ford	Woody	107th Evac. Hosp.	
Forsythe	Jim	106th Inf. Div. 424th Regiment	A Co
Frank	Dick	80th Inf. Div. 317th Regiment	
Fraser	D.W.	517th Prcht Inf.	
Freeman	Terrence	RNVR	DEMS
French	Arthur	740th Tank Bn	H Co
Garrett	Frank	99th Inf. Div. 393rd Regiment	Service Battery
Gatens	John	106th Inf. Div. 589th Bn	A Battery
Gaudere	Francis	30th Inf. Div. 119th Regiment	1st Bn
Gembel	John	1st Inf. Div. 26th Regiment	D Co
Genthner	Ed	16th Arm. Div. 18th MIB	

Glamore	James	1st Inf. Div. 26th Regiment	C Co
Goldstein	Al	99th Inf. Div. 924th FA	
Goldston	Johnny	5520th AAA (AW)MLB	HQ Co
Gonzales	H.A. Red	78th Inf. Div. 311th Regiment	M Co
Greene	Vernon	78th Inf. Div. 311th Regiment	M Co
Grubb	Clyde	3rd Arm. Div. 36th Inf. Regiment	A Co
Gundy	Ted	99th Inf. Div. 393rd Regiment	B Co
Hale	Howard	Tank Bn	32nd Rcn Mech
Hammerlund	Roy	75th Inf. Div. 290th Regiment	K Co
Hanson	Arlow	1st Inf. Div. 16th Regiment	
Hartman	David	CBI – OSS	
Haskett	Chuck	8th AF 487th Bomb Gr.	
Hathaway	Jack	1st Inf. Div. 26th Regiment	D Co
Havens	Bob	75th Inf. Div. 291st Regiment	HQ Co
Hawkins	Arthur	75th Inf. Div. 290th Regiment	I Co
Head	Woodrow	1st Inf. Div. 26th Regiment	D Co
Henderson	B.C.	99th Inf. Div. 394th Regiment	B Co
Hennessey	Paul	78th Inf. Div. 309th Regiment	
Henry	Richard	8th AF 381st HB Gr.	
Herring	Jess	78th Inf. Div. 309th Regiment	L Co
Hicks	H.	83rd Inf. Div. 330th Regiment	C Co
Hill	O.B.	82nd AB 508 PIR	
Hinchman	Jack	1st Inf. Div. 26th Regiment	D Co
Hoff	Russell	106th Inf. Div. 422nd Regiment	M Co
Hojnowski	Ed	99th Inf. Div. 324 Med. Bn	
Holm	Bob	1st Inf. Div. 16th Regiment	HQ Co
Holster	Bob	9th Arm. Div. 9th Engr. Bn	A Co
Houseman	Don	106th Inf. Div. 423rd Regiment	D Co
Hoye	John	316th T.C. Group	
Hubbard	David	Signal Section	HQ Co
Hummingbird	Raymond	740th Tank Bn	HQ Co
Hunt	Don	U.S. Navy	
Hunter	Johnnie	3218th AM Supply	A Co
Hutchings	Elby	740th Tank Bn	Service Co
Ince	Ken	113th ARV AAAF	Charter Towers
Ingram	Tom	90th Inf. Div. 359th Regiment	F Co
Ingram	John	99th Inf. Div. 394th Regiment	E Co
Jacobson	Arthur	83rd Inf. Div. 330th Regiment	I Co
Johnson	Maury	87th Inf. Div. 345th Regiment	AT Co & L Co
Johnson	Darrell	83rd AB 551 P.I.R.	

Jundanian	Tom	99th Inf. Div. 394th Regiment	Cannon Co
Kast	Rolland	99th Inf. Div. 395th Regiment	I Co
Keaton	Frank	30th Inf. Div.	
Keen	J.D.	740th Tank Bn	C Co
Kellen	Bob	1st Inf. Div. + 99th Inf. Div.	
Kimble	Billy	740th Tank Bn	A Co
Kirkendall	Hope	Army Nurse	
Kisser	Jim	78th Inf. Div. 552nd AAA	HQ Co
Kline	Morris	84th Inf. Div. 335th Regiment	
Kowalski	Lee	1st Inf. Div. 26th Regiment	C Co
Kraemer	Ron	99th Inf. Div. 394th Regiment	D Co
Kristan	Albert	3548 Ord.	MAM Co
Kurzeja	Mike	106th Inf. Div. 423rd Regiment	H Co
Kvingedal	Bjarne	99th Inf. Battalion	
Laisure	Jack	99th Inf. Div. 324th Engineer	HQ Co
Lamson	Ernie	82nd AB 508th PIR	A Co
Landry	Joe	776th AAA AWBM	3rd Army
Lange	Jerry	740th Tank Bn	A Co
Lebda	John	1st Inf. Div. 26th Regiment	D Co
Lee	William	1st Inf. Div. 26th Regiment	D Co
Leh	Ernest	1st Inf. Div. 18th Regiment	E Co
Leopold	William	75th Inf. Div. 291st Regiment	C Co
Lester	David	30rd Inf. Div.	
Levitsky	Dorothy	164th General Hospital	
Levitsky	Ellan	164th General Hospital	
Levy	Dalton	29th Inf. Div. 175th Regiment	C Co
Lombardo	Samuel	99th Inf. Div. 394th Regiment	I Co
Lovelace	Earl	2nd Inf. Div. 38th Regiment	M Co
Lowenstein	Robert	299th Cmbt. Eng. Bn	
Lytle	Glenn	104th Inf. Div. 413th Regiment	K Co
Mackey	John	754 FA A Battery	
Madsen	Fred	30th Inf. Div. 120th Regiment	B Co
Maher	Bill	1st Inf. Div. 26th Regiment	M Co
Maresca	Frank	75th Inf. Div. 289th Regiment	F Co
Mastrangelo	Vito	607th Graves Regiment	
Mayrsohn	Barney	106th Inf. Div.	
McAuliffe	John	87th Inf. Div. 347th Regiment	M Co
McCracken	Harry	99th Inf. Div. 395th Regiment	Medic
McGowan	Chuck	99th Inf. Div. 395th Regiment	C Co
McIlroy	J.R.	99th Inf. Div. 393rd Regiment	F Co

McLemore	John	1st Inf. Div. 26th Regiment	F Co
McMullen	Fred	1st Inf. Div. 18th Regiment	Anti-tank Co
McMurdie	Bill	99th Inf. Div. 394th Regiment	A Co
Mejia	Juan	106th Inf. Div. 424th Regiment	L Co
Merrill	Mazel	740th Tank Bn	A Co
Messina	Sam	1st Inf. Div. 26th Regiment	C Co
Meyer	George	1655 Engr.	
Meyer	Bill	99th Inf. Div. 371st FA	Service Battery
Middleton	J.P.	99th Inf. Div. 393rd Regiment	G Co
Miller	Franklin	740th Tank Bn	Service Co
Miller	Harry	740th Tank Bn	HQ Co
Miller	Frank	99th Signal Corps	
Miller	William	558th AAA (AW)	HQ Bn
Mitcherson	Audrey	99th Inf. Div. 394th Regiment	D Co
Moran	Jack	87th Inf. Div. 347th Regiment	K Co
Morehouse	Philip	1st Inf. Div.	HQ Co
Morettini	Joseph	82nd Airb. 508th Regiment	E Co
Moretto	Rocky	1st Inf. Div. 26th Regiment	C Co
Morphis	Bert	1st Inf. Div. 26th Regiment	B Co
Morris	Don	1st Inf. Div.26th Regiment	HQ Co
Mozden	Joe	99th Inf. Div. 394th Regiment	L Co
Muggar	Victor	1st Inf. Div. 26th Regiment	C Co
Munford	Tom	740th Tank Bn	A Co
Najarian	Helen	Army Nurse	
Nauss	Lovern	1st Inf. Div. 1st Sign. Cps	
Nelson	Al	99th Inf. Div. 393rd Regiment	I Co
Nelson	Ruth	Army Nurse	
Nolle	Bill	75th Inf. Div. 291st Bn	C Co
Nordgreen	Robert	83rd Inf. Div. 329th Regiment	K Co
O'Hare	John	30th Inf. Div. 117th Reg	E Co
Paluch	Ted	285th FA OB	
Parker	John	8th AF 457th Bomb Gr.	
Parker	Fred	1st Inf. Div. 26th Regiment	
Parsons	Bill	78th Inf. Div. 311th Regiment	D Co
Pashuck	Walter	57th Field Hospital	
Patlen	Bernard	106th Inf. Div. 424th Regiment	E Co
Patterson	Harold	86th Chim. Mortar Bn	
Pearson	Paul	740th Tank Bn	C Co
Peckham	Alden	1st Inf. Div. 26th Regiment	C Co
Peiffer	Warren	99th Inf. Div. 371st FA	Service Battery

Peltier	Gordon	1st Inf. Div. 33 FA	
Pickett	Walter	83rd Inf. Div. 908 FA Bn	
Pidcoe	Robert	1st Inf. Div. 26th Regiment	HQ Co
Piergiovanni	Pete	17th Airborne	
Pietroforte	Joseph	1st Inf. Div. 18th Regiment	2nd Bn
Pildner	John	75th Inf. Div. 290th Regiment	Anti-tank Co
Preston	Al	1st Inf. Div. 26th Regiment	
Price	Franck	1st Inf. Div. 16th Regiment	M Co
Pulver	Murray	30th Inf. Div. 120th Regiment	B Co
Ramaly	Clayton	1st Inf. Div. 26th Regiment	D Co
Randall	Fred	1st Inf. Div. 18th Regiment	
Reid	Paul	1st Inf. Div. 26th Regiment	G Co
Reno	Clarence	9th Arm. Div.	
Rice	Darold	90th Inf. Div. 359th Regiment	D Co
Rich	Sheldon	78th Inf. Div.	778th Ord Co
Roach	Robert	78th Inf. Div. 310th Regiment	I Co
Rogers	Hope	16th General Hospital	
Ronningen	Thor	99th Inf. Div. 395th Regiment	I Co
Rubel	George	740th Tank Bn	
Russ	Matt	12th Army Group	
Ruth	Bill	3rd Arm. Div. 33rd Regiment	
Ryan	Don	9th Army 787th FA Bn	
Ryan	Bill	1st Inf. Div. 16th Regiment	
Sackett	Bob	1st Inf. Div. Band	
Samuelson	Earl	99th Inf. Div. 395th Regiment	G Co
Sassin	Richard	75th Inf. Div.	
Savage	Clifford	99th Inf. Div. 393rd Regiment	M Co
Savage	Job	558th AAA (AW)Bn	Battery D
Savard	John	2nd Inf. Div. 38th Regiment	G Co
Schaefer	Harold	99th Inf. Div. 394th Regiment	G Co
Schaffner	John	106th Inf. Div. 589 FA Bn	A Battery
Schmidt	Gerald	75th Inf. Div.	C Co
Scholtz	Ralph	69th Inf. Div. 273rd Regiment	HQ Co
Schott	Joseph	U.S. Navy	
Schumacher	Paul	9th Inf. Div. 39th Regiment	C Co
Scoggins	Lloyd	740th Tank Bn	B Co
Seidenstricker	Arthur	69th Inf. Div. 273rd Regiment	HQ Co
Shea	Tom	1st Inf. Div. 26th Regiment	C Co
Shilliday	Ted	99th Inf. Div. 393rd Regiment	
Simmons	Buster	30th Inf. Div. 120th Regiment	Med. Det.

Smith	Mike	740th Tank Bn	C Co
Smith	Ken	Royal Navy	Light Coastal Forces
Smith	Ken	106th Inf. Div. 423 Regiment	H Co
Smolens	Irwing	4th Inf. Div. 29th FA	B Battery
Sniezak	Al	375th Med. Bn	C Co
Sorenson	Arthur	104th Inf. Div. 414th Regiment	L Co
Soucie	Earl	552nd AAA (AW) Bn	C Battery
Sprague	Carlton	558th AAA (AW) Bn	Battery C
Stanger	Kenneth	99th Inf. Div. 395th Regiment	Cannon Co
Stefanavage	Pete	1st Inf. Div. 26th Regiment	
Steinsiek	Walter	U.S. Merchant Marines	
Strebel	Frank	104th Inf. Div. 413th Regiment	F Co
Strong	George	106th Inf. Div. 423rd Regiment	HQ Co
Stumpff	Del	99th Inf. Div. 394th Regiment	D Co
Stumpff	Del	99th Inf. Div. 394th Regiment	D Co
Swanson	V.E.	99th Inf. Div. 395th Regiment	C Co
Swengel	John	78th Inf. Div. 309th Regiment	L Co
Swett	John	106th Inf. Div. 423rd Regiment	H Co
Swift	Ed	75th Inf. Div.	
Tafoya	Ernesto	U.S. Navy	
Tanner	Douglas	740th Tank Bn	Service Co
Taylor	Arnold	99th Inf. Div.	MP
Tervooren	Joseph	740th Tank Bn	Service Co
Thomas	Bob	87th Inf. Div. 347th Regiment	I Co
Thomas	James	740th Tank Bn	Service Co
Tolmasov	James	99th Inf. Div. 395th Regiment	L Co
Tolstedt	Grandon	104th Inf. Div. 414th Regiment	A Co
Towers	Frank	30th Inf. Div. 117th Regiment	
Trumbly	Bill	2nd Inf. Div. 38th Regiment	
Tullier	John	740th Tank Bn	HQ Co
Tunis	Joe	Solomon, Pacific	
Turkington	Martin	75th Inf. Div. 289th Regiment	B Co
Tylka	Joseph	99th Inf. Div. 395th Regiment	L Co
Van Osdol	Bill	U.S. Navy	
Vansweden	Lee	U.S. Navy	
Vargo	Frank	1st Inf. Div. 18th Regiment	L Co
Vedeloff	Russell	75th Inf. Div. 290th Regiment	H Co
Vervack	Paul	1st Inf. Div. 26th Regiment	A Co
Vinson	A.J.	740th Tank Bn	C Co
Vogel	Bernard	78th Inf. Div. 310th Can.	

Wagner	Harry	1st Inf. Div. 26th Regiment	C Co
Walcott	Kenneth	740th Tank Bn	Service Co
Ward	Leonard	107th Engr « C Bn	254 HQ Co
Warsaw	Ernie	100th Bomb Gr 418th Sq	
Webster	C.O.	740th Tank Bn	A Co
Weed	Don	78th Inf. Div. 310th Regiment	D Co
Weisenberg	Gene	U.S. Navy	
Weisenberg	Joe	U.S. Navy	
Weisenberg	Jack	U.S. Navy	
Weiss	Herman	324th Engr.	C Co
Weissman	Morris	1867th Labor Super CO	Henri-Chapelle
Wenc	Chester	106th Inf. Div. 424th Reg	B Co
Werme	Russell	1st Inf. Div. 26th Regiment	C Co
Whitehead	Charles	99th Inf. Div. 370th FA Bn	
Whitmarsh	B.A.	99th Inf. Div. 395th Regiment	C Co
Wiberg	Don	99th Inf. Div. 394th Regiment	A Co
Wilkins	B.O.	99th Inf. Div. 393rd Regiment	K Co
Wilkinson	Jerry	Trucking Co	
Williams	Joseph	4049 QM Trk Co	
Woelkers	Harry	28th Inf. Div. 28th Rcn.	
Wojtusik	Stanley	106th Inf. Div. 422nd Regiment	G Co
Wolf	Carl	1st Inf. Div. 16th Regiment	I Co
Woodman	Charlie	75th Inf. Div. 291st Regiment	B Co
Woodrow	Charles	99th Inf. Div. 924th FA	Signal Corps
Wright	Lloyd	740th Tank Bn	A Co
Yahn	Orville	1st Inf. Div. 26th Regiment	C Co
Zeece	Steve	20th Air Force	
Zimmerman	Keith	104th Inf. Div.	HQ Co
Zink	Charles	78th Inf. Div. 309th FA Bn	B Battery
Zuskin	Bennie	1st Inf. Div. 26th Regiment	C Co

APPENDIX C

SOURCES

607th Quartermaster Graves Registration Company Unit History, WWII US Medical Research Centre Website, https://www.med-dept.com/unit-histories/607th-quartermaster-graves-registration-company

Aalberts, Georges: Personal Interview by Author, November 2021.

Amero, Rick: Personal Communications with Author re: Charles Amero.

Belgium, The Official Account of What Happened 1939-1940, Evans Brothers 1st Edition, Ltd., 1941.

Clements, Philip: Personal Communications with Author re: William Clements.

Clements, William: Interview with New York State Military Museum, February 7, 2002; Interview with *The Saratogian*, October 15, 2002.

Coffman, Warren: *I Never Intended to be a Soldier*, Lifestyles Press, 2000.

Didden, Alice: Personal Interview, November 2021.

Gendron, Thomas: *The Operations of the 2nd Battalion, 26th Infantry at Dom Butgenbach, Belgium, 18-21 December, 1944, Advanced Infantry Officers Course,* Fort Benning, GA, 1949.

Lebda, John F.: *A Million Miles to Go*, Xlibris US, 2010

Mastrangelo, Vito: Personal Interview by Author, October 2022; Fox News Interview, June 6, 2019.

Moretto, Jill: Personal Communications re: Rocco Moretto

Moretto, Rocco (Rocky): *D-Day Experience*, Written for Professor Ambrose, University of New Orleans, June 16, 1989; *A Memorable Experience*, Written for the Battle of the Bulge History Book, date unknown; *A WWII Infantryman's Experience Across Europe*, date unknown.

Putters, Charles: Personal Interview by Author, November 2021.

Rogester, Collette: Personal Interview and Communications with Author, January 2022.

Schmetz, Marcel: Personal Interviews and Communications with Author, Various dates.

Schmetz, Mathilde: Personal Interviews and Communications with Author, Various dates.

September 1944 – September 1999, 55e anniversaire de la Liberation, 5e anniversaire du Remember Museum '39-'45, Imprimerie Fortemps, 1999.

Souvenirs de la guerre, 1940-1945 á Thimister-Clermont, Centre Culturel, 2005.

Wijers, Hans: *Battle of the Bulge, Volume II, Hell at Bütgenbach/Seize the Bridges*, Stackpole Books, 2010.

www.ingramcontent.com/pod-product-compliance
Lightning Source LLC
LaVergne TN
LVHW010658110826
845149LV00014B/3150

* 9 7 9 8 9 8 9 3 5 8 5 0 2 *